IEE TOPICS IN CONTROL SERIES 3
Series Editors: Prof. B.H. Swanick
Prof. H. Nicholson

State-space and frequency-domain methods in the control of distributed parameter systems

Previous volumes in this series

Volume 1 Modelling and simulation
J.R. Leigh

Volume 2 Essentials of nonlinear control theory
J.R. Leigh

State-space and frequency-domain methods in the control of distributed parameter systems

Stephen P. Banks

PETER PEREGRINUS LTD
on behalf of the
Institution of Electrical Engineers

Published by Peter Peregrinus Ltd., London, UK.

©1983: Peter Peregrinus Ltd

All rights reserved. No part of this publication may be reproduced, stored in a retrieval system or transmitted in any form or by any means—electronic, mechanical, photocopying, recording or otherwise—without the prior written permission of the publisher.

British Library Cataloguing in Publication Data

Banks, S.P.
State-space and frequency-domain methods in the control of distributed parameter systems. —(IEE topics in control series; 3)
1. Distributed parameter systems—Mathematical models
I. Title II. Series
629.8 QA402

ISBN 0-86341-000-6

Printed in England by Short Run Press Ltd., Exeter

To Yvonne and David

Contents

Preface

This book is an exposition of some of the main areas of distributed parameter control and contains chapters on stability, controllability and realisation theory, optimal control and the distributed root locus. The main purpose is to present the theory of systems described by partial differential equations, although other types of systems are also discussed.

The book should be suitable for a graduate level course on the control of distributed parameter systems, and takes the student to some of the current areas of research in this field. There are many topics which could not be included in a book of this length; for example, there is little mention of the control of nonlinear systems. However, this is still a fairly new area in distributed control theory and its inclusion was not felt to be appropriate in the present book.

Theorems, definitions, etc. are numbered consecutively within each section; e.g. theorem 3.4.2 may follow definition 3.4.1. The references are given by the Harvard System and can be found at the end of the book.

I would like to thank Professor H. Nicholson for suggesting that I should write a book in this area and Professor A. J. Pritchard and Dr. R. F. Curtain who introduced me to this subject while I was at Warwick University.

S. P. Banks
Sheffield, April 1983

Chapter 1

Introduction

1.1 Control theory

The theory of control can be traced back to its origins in electronic circuit theory, for example, in the design of amplifiers which are stable to temperature variations, by using resistive feedback. These ideas were generalised by Nyquist (1932) and Bode (1954) into techniques which could be used to stabilise any finite dimensional single-input single-output system (which can be defined by a rational transfer function). This led to the classical theory of compensation and root locus methods for such systems, as presented, for example, in Gupta and Hasdorff (1970).

The techniques developed for linear finite dimensional systems with a single input and a single output are very effective and widely applied. However, when the system is nonlinear, or when we wish to consider some kind of optimum control, then it is usually more convenient to use a state-space approach. In other words, we consider the system to be defined by a finite dimensional differential (or difference) equation. This theory was stimulated from various directions – minimum time or energy control of space vehicles, economic theory, aircraft flight controllers etc. – and is discussed in detail in the text of Lee and Markus (1967). The general theory of nonlinear optimisation is given in Pontryagin *et al.* (1962) and the theory of stability originated in the work of Lyapunov (1949). The question of stability of a system is, of course, central to control theory and is now a highly developed subject in its own right (see, for example, Hahn, 1963; LaSalle and Lefschetz, 1961; Yoshizawa, 1966; and Halanay, 1966).

Interest in the s-domain approach to control was again stimulated by the desire to control systems with many inputs and outputs (i.e. multivariable systems). Computer-aided design of such systems was developed by Rosenbrock (1974) and the general theory of compensators and the root locus theory by Wolovich (1974), Postlethwaite and MacFarlane (1979) and Owens (1978). These studies highlighted the need for precise definitions of concepts such as system poles and zeros and related the state-space theory to the

s-domain formulation by realisation theory. This is now highly developed and awaits only to be widely applied in the industrial environment.

The above work is, however, entirely concerned with lumped systems. In reality, all systems are distributed and are described more accurately by partial differential, integral or delay equations. It is therefore clear that a more general theory embracing all such systems is necessary. The state-space approach to distributed systems has been vigorously developed in the last decade and the linear theory can now be said to be fairly complete. We shall present the essential aspects of this theory in this book. However, just as was the case with multivariable systems, the s-domain methods are only just beginning to be studied in detail; we shall discuss these techniques, and in particular the distributed root locus, in the last chapter. The general theory of the root locus of systems defined by unbounded operators remains an open problem.

1.2 Methods of analysis

The basic ideas involved in studying infinite dimensional systems are similar to those which have been used in finite dimensional theory. The main difference is in the mathematical technicalities which must be overcome before a complete understanding of the theory can be gained. For this reason, Chapter 2 will present (mostly without proof) the mathematical ideas which are needed in this book. They consist largely of results from functional analysis and a thorough study of this chapter is strongly recommended.

The simplest infinite dimensional systems are those given by an equation of the form

$$\dot{x} = Ax + bu$$

where A is a compact operator or has compact resolvent (see Chapter 2 for definitions). The main reason for this is that the spectrum of such operators is disjoint (apart from in the former case which may have an accumulation point at the origin) and we may find 'good' finite dimensional approximations to these equations. In this sense the results obtained for such systems are perhaps not particularly surprising. However, for more general systems we cannot expect to obtain as precise results as we can for finite dimensional systems. Indeed, the theory of stability for general systems has some surprises (see Chapter 4), which could not be predicted by using finite dimensional ideas.

One of the most interesting areas of infinite dimensional systems theory is in the application of spectral theory to the root locus of distributed systems. This will be considered in the final chapter, mainly for systems with a bounded realisation. By correctly interpreting the transfer function (which may be irrational) for such a system, we shall show that many of the characteristics of the classical root locus can be generalised to infinite dimensional systems.

1.3 Summary

We shall finish this chapter with a summary of the ideas which follow. In the second chapter the basic results of functional analysis are presented mainly without proof. The proofs of these results may be found, for example, in Yosida (1974), Dunford and Schwartz (1959), Kato (1976) or Taylor (1958). The mathematics contained in Chapter 2 is the basis for all that is to follow.

In Chapter 3 the theory of infinite dimensional differential equations is discussed from the semigroup viewpoint and a number of system examples is given which will be used later in the book. The semigroup for a system of the form

$$\dot{x} = Ax$$

where A is some operator on a Banach space is just the operator which corresponds to the solution of the system through x_0 at $t = 0$, and is the infinite dimensional counterpart of the matrix e^{At} for finite dimensional systems. The examples which we discuss in Chapter 3 are taken from various types of physical systems. These will include the heat and wave equations, delay equations, plasma dynamic equations and a simple nonlinear nuclear reactor model. There are, of course, an almost unlimited variety of such examples, but it is hoped that the ones chosen sufficiently represent those which are of interest.

The theory of stability of distributed systems is given in Chapter 4. The main results are the general Lyapunov theory for nonlinear systems, the theory of linear systems (in particular, the generalisation of the classical spectral theory) and the input–output theory (including the circle theorem). It will be seen that for systems with compact resolvent, the stability of linear systems is closely related to the finite dimensional results.

Chapter 5 is devoted to the theory of controllability, observability, stabilisability and realisation theory. The first three concepts are very similar to the corresponding finite dimensional ones. In realisation theory we shall see that there is not necessarily such a close relation between the system operator and the singularities of the transfer function as in lumped systems.

The theory of optimal control is an important area for applications and in Chapter 6 we shall study the linear-quadratic problem, the receding horizon solution and the time optimal problem. The receding horizon solution is a nonlinear feedback law which improves on the linear-quadratic solution when small noise perturbations are present in the system.

In the final chapter we shall consider the s-domain approach to infinite dimensional systems theory. This theory is still being developed and so this chapter will be of a more expository nature than the earlier ones and the main concern will be with the root locus of bounded systems. This will involve an application of realisation theory and the spectral decomposition of operators.

Chapter 2

Mathematical prerequisites

2.1 Introduction

The control theory of lumped-parameter systems depends to a large extent on a knowledge of matrices and differential equations, and as such is mathematically elementary. However, distributed systems are governed by delay, partial differential or integro-differential equations and are described naturally on spaces of infinite dimension. The mathematical requirements for such systems are therefore both algebraic and topological and one must consider system states which are not just points (in Euclidean space) but consist of whole functions. We shall, therefore, present in this chapter the aspects of functional analysis which are needed for our purposes; of course, in a book of this length only an outline of the subject can be given, and the reader should consult the literature for a more complete exposition. We shall assume that the reader is familiar with undergraduate algebra and analysis and start immediately with the concepts of Hilbert and Banach spaces.

2.2 Elements of Hilbert and Banach space theory

Definition 2.2.1
A real (or complex) *Banach Space* X is a complete normed vector space over $\mathscr{R}$ (or $\mathscr{C}$). Recall that the norm satisfies the following axioms:

(a) $\|x\| = 0$ iff $x = 0$
(b) $\|\alpha x\| = |\alpha| \, \|x\|$, $\forall \, \alpha \in F, x \in X$
(c) $\|x + y\| \leq \|x\| + \|y\|$, $\forall \, x, y \in X$

where F denotes either $\mathscr{R}$ or $\mathscr{C}$.

Definition 2.2.2
A (real or complex) *Hilbert Space* H is a Banach space whose norm is defined

by an inner product, denoted by $\langle\cdot,\cdot\rangle$ (or $\langle\cdot,\cdot\rangle_H$, when several Hilbert spaces are being considered).

Because of the lack of space, the following basic results concerning the structure of Hilbert spaces will be presented without proof (see Yosida, 1974; Horvath, 1966; Taylor, 1958). Also, since we shall require spaces of at most countable dimension, we shall restrict attention to this case.

Proposition 2.2.3 (parallelogram law)
We have, in any Hilbert space H,

$$\|x + y\|^2 + \|x - y\|^2 = 2\|x\|^2 + 2\|y\|^2, \ \forall\, x, y \in H. \quad \square$$

(without further comment, H will denote a Hilbert space).

Proposition 2.2.4
If $A \subseteq H$ is closed and convex and $h \in H$, then $\exists$ a unique $x \in A$ such that $\|x - h\| \leqslant \|y - h\|$, $\forall\, y \in A$. $\square$

Recall that subsets $S_1, S_2 \subseteq H$ are orthogonal (written $S_1 \perp S_2$) if $\langle h_1, h_2\rangle = 0$, $\forall h_1 \in S_1$, $h_2 \in S_2$. Then it is easy to see that if $E \subseteq H$ is closed, every $h \in H$ has the unique representation $h = f + f_\perp$ where $f \in E$ and $f_\perp \perp E$.

Definition 2.2.5
A set $\{e_i\}_{i \in I} \subseteq H$ is *orthonormal* if $\langle e_i, e_j\rangle = \delta_{ij}$, $\forall i, j \in I$. This set is *maximal* if it is not properly contained in any other orthonormal set.

If $S \subseteq H$ is any subset, let $\overline{G(S)}$ denote the smallest closed subspace of H containing S. It follows easily from the above remarks that every Hilbert space contains a maximal orthonormal family $\{e_i\}_{i \in I}$ such that

$$H = \overline{G(\{e_i\}_{i \in I})}$$

Such an orthonormal family is called a *basis* of H. If I is countable, H is said to be *separable*. We now restrict attention to separable spaces.

Theorem 2.2.6
If H is separable, we have:

(a) $$\sum_{i=1}^{k} |\langle x, e_i\rangle|^2 \leqslant \|x\|^2, \ \forall\, x \in H \qquad \text{(Bessel's inequality)} \tag{2.1}$$

for any orthonormal family $\{e_i\} \subseteq H$ (possibly with $k = \infty$).
(b) If $\{e_i\}$ is a basis, then

$$x = \sum_{i=1}^{\infty} \langle x, e_i\rangle e_i$$

and

$$\|x\|^2 = \sum_{i=1}^{\infty} |\langle x, e_i\rangle|^2 \qquad \text{(Parseval's relation)} \tag{2.2}$$

Definition 2.2.7
A (continuous) *linear form* (or linear functional) x^* on a Banach space X is a continuous linear map $x^*:X \to F\ (=\mathcal{R}$ or $\mathcal{C})$. The *dual space* X^* of X is the set of all linear forms on X. X is called *reflexive* if the map $J:X \to X^{**}\ (=(X^*)^*)$ defined by

$$\langle J(x), x^* \rangle = \langle x, x^* \rangle$$

is onto X^{**}. Note that we use the notation $\langle x, x^* \rangle\ (=\langle x^*, x \rangle) \triangleq x^*(x)$ if $x^* \in X^*$, $x \in X$.

X^* becomes a Banach space under the norm

$$\| x^* \| = \sup_{\|x\|=1} |\langle x^*, x \rangle| \tag{2.3}$$

and then J is an isometric isomorphism (in general, into) X^{**}. If H is a Hilbert space we have even more.

Proposition 2.2.8
If $h^* \in H^*$, then $\exists$ a unique $h \in H$ such that

$$\langle h^*, x \rangle_{H^*,H} = \langle x, h \rangle_H, \ \forall x \in H$$

and $\|h^*\| = \|h\|$. □ (when we need to be explicit about dualities, we shall add subscripts as above).

It follows that H is isometrically (semi-linearly) isomorphic to H^*, but it is important to note that, whereas the map J above is canonical, the map $j_H:h \to h^*$ (given by proposition 2.2.8) is not. We often consider the case of two Hilbert spaces V and H where V is a dense subspace of H and such that the map $i_V:V \subseteq H$ (natural injection) is continuous, i.e. $\exists$ a constant $c > 0$ such that

$$\|h\|_H \leqslant c \| h \|_V$$

Then, by using the map j_H we have the sequence

$$\overset{i_V}{V \subseteq} \overset{j_H}{H \subseteq} \overset{i_{H^*}}{H^* \subseteq} V^* \tag{2.4}$$

Students are frequently confused by this structure since we know that V is isometrically isomorphic to V^*, by the map j_V. However, we should note that the combined map $i_{H^*} \circ j_H \circ i_V$ in (2.4) is not j_V. (Note that the last inclusion in (2.4) makes sense since if $h^* \in H^*$ then $h^*|_V \subset V^*$, by the continuity of i_V. If $h_1^*, h_2^* \in H^*$ are such that $h_1^*|_V = h_2^*|_V$ then $h_1^* - h_2^* = 0$ on V and by continuity and density of V in H, $h_1^* = h_2^*$ in H, i.e., i_{H^*} is an injection.)

Examples 2.2.9
(a) Let $\Omega \subseteq \mathcal{R}^n$ be open and bounded and let $C^k(\bar{\Omega})$, $k \in N_+$ be the set of all functions defined on Ω whose derivatives to order k exist and are uniformly

continuous, and define a norm on $C^k(\bar{\Omega})$ by

$$\|f\|_k \triangleq \sup_{|p|\leqslant k} \{\sup_{x\in\Omega} |(\partial/\partial x)^p f(x)|\}, f \in C^k(\bar{\Omega}) \tag{2.5}$$

where $p \in N^n$, $|p| = p_1 + \ldots + p_n$ and $(\partial/\partial x)^p \triangleq (\partial/\partial x_1)^{p_1} \ldots (\partial/\partial x_n)^{p_n}$. Then $C^k(\bar{\Omega})$ is a Banach space. We write $C(\bar{\Omega}) \triangleq C^0(\bar{\Omega})$ and define $C^\infty(\bar{\Omega}) = \bigcap_{k=0}^{\infty} C^k(\bar{\Omega})$. Note that $C^k(\bar{\Omega})$ is not a Hilbert space.

(b) Let $\Omega \subseteq \mathcal{R}^n$ and define the spaces $L^p(\Omega)$, $p \in \mathcal{R}_+$, as the sets of (Lebesgue) measurable functions on Ω such that

$$\|f\|_{L^p(\Omega)} \triangleq (\int |f(x)|^p \mathrm{d}x)^{1/p} < \infty, p < \infty \tag{2.6}$$

$$\|f\|_{L^\infty(\Omega)} \triangleq \operatorname*{ess\,sup}_{x\in\Omega} |f(x)| < \infty \tag{2.7}$$

(Recall that functions equal almost everywhere are identified.) If $1 \leqslant p \leqslant \infty$, then $L^p(\Omega)$ is a Banach space with the above norm, and, moreover, $L^2(\Omega)$ is a Hilbert space with the inner product

$$\langle f, g\rangle_{L^2(\Omega)} = \int_\Omega f(x)\bar{g}(x)\mathrm{d}x \tag{2.8}$$

(c) Let l^p, $1 \leqslant p \leqslant \infty$ be the spaces of sequences $x = \{x_i\}_{i=1,2,\ldots}$ of elements of F such that

$$\|x\|_{l^p} \triangleq \left(\sum_{i=1}^{\infty} |x_i|^p\right)^{1/p} < \infty, \qquad 1 \leqslant p < \infty \tag{2.9}$$

$$\|x\|_{l^\infty} \triangleq \sup_{1\leqslant i<\infty} |x_i| < \infty, \qquad p = \infty \tag{2.10}$$

l^p is again a Banach space and l^2 is a Hilbert space

Proposition 2.2.10 (Lax-Milgram)
Let B be a continuous sesquilinear form on a Hilbert space H (i.e. $B : H \times H \to F$ is linear in the first argument and semilinear in the second) such that $|B(x,y)| \geqslant m\|x\|^2$, $\forall\, x \in H$, $m > 0$, and let $f \in H^*$. Then $\exists$ a unique $h \in H$ such that $f(x) = B(x,h)\ \forall x \in H$. □

Theorem 2.2.11 (Banach-Steinhaus)
Suppose that $(x_i^*)_{i\in I} \subseteq X$ (a Banach space) and that the sequence $\{\langle x_i^*, x\rangle\}_{i\in I}$ is is bounded for each $x \in X$, then the sequence $\{\|x_i^*\|\}_{i\in I}$ is also bounded. □

Theorem 2.2.12 (Hahn-Banach)
Let $M \subsetneqq X$ be a subspace. If $m^* \in M^*$, then $\exists\, x^* \in X^*$ such that $\|x^*\| = \|m^*\|$ and

$$x^*(x) = m^*(x) \quad \text{for } x \in M. \square$$

Before leaving this section mention should be made of the notion of quotient space. If X is a Banach space and M is a closed subspace, define the vector space

$$\tilde{X} \triangleq X/M = \{\tilde{x} : \tilde{x} = x + M\} \tag{2.11}$$

i.e. $\tilde{X}$ is the set of all affine subspaces parallel to M. Defining

$$\|\tilde{x}\| = \inf_{y \in \tilde{x}} \|y\| = \inf_{m \in M} \|x - m\| = \operatorname{dist}(x, M)\,, \tag{2.12}$$

$\tilde{X}$ becomes a Banach space.

2.3 Operator Theory

Definition 2.3.1
Let X and Y be Banach spaces and let A be a function from a linear manifold $\mathscr{D}(A)$ of X into Y such that

$$A(\alpha x_1 + \beta x_2) = \alpha A x_1 + \beta A x_2, x_1, x_2 \in \mathscr{D}(A), \alpha, \beta \in F \tag{2.13}$$

Then A is a *linear operator* with domain $\mathscr{D}(A)$. The *kernel* of A is defined by

$$\ker A = \{x \in \mathscr{D}(A) : Ax = 0\} \tag{2.14}$$

A is one-to-one iff $\ker A = 0$. In the latter case we can define the operator A^{-1} (the inverse of A) with domain $\mathscr{R}(A)$ (the range of A).

If B is another linear operator such that $\mathscr{D}(B) \subseteq \mathscr{D}(A)$ and

$$Bx = Ax, \quad \forall x \in \mathscr{D}(B)$$

we say that A is an *extension* of B and we write $A \supseteq B$ (or $B \subseteq A$). Consider now the case where $\mathscr{D}(A) = X$ and

$$\|A\| \stackrel{\Delta}{=} \sup_{\|x\|=1} \|Ax\| \; (= \sup_{x \neq 0} \|Ax\|/\|x\|) < \infty \tag{2.15}$$

Then A is continuous (in the metric topologies of X and Y) and we say that A is a *bounded operator*. The space of all bounded operators defined on X (into Y) is denoted by $\mathscr{B}(X, Y)$ (or $\mathscr{L}(X, Y)$), (or $\mathscr{B}(X)$ if $X = Y$) and it is easy to see that $\mathscr{B}(X, Y)$ is a Banach space (with the obvious vector space structure and the norm (2.15)) and since, when $X = Y$, we can define a product $A_1 A_2$ of operators on X and $\|A_1 A_2\| \leq \|A_1\| \, \|A_2\|$ (by 2.15)). It follows that $\mathscr{B}(X)$ is a Banach algebra.

Theorem 2.3.2
If X is a Banach space and $A \in \mathscr{B}(X)$ with $\|A\| < 1$, then $(I - A)^{-1}$ exists (in $\mathscr{B}(X)$) and

$$(I - A)^{-1} = I + A + A^2 + \ldots \quad \text{(Neumann series)} \tag{2.16}$$

with convergence in the $\mathscr{B}(X)$ topology.□

A particularly important class of operators is the class of projections.

Definition 2.3.3
An operator $P \in \mathscr{B}(X)$ is a *projection* (on $\mathscr{R}(P)$) if $P^2 = P$.

If H is a Hilbert space and M is a closed subspace, then

$$H = M \oplus M^{\perp} \tag{2.17}$$

where $M^{\perp} = \{h \in H : \langle h, m \rangle = 0, \forall m \in M\}$ (the *orthogonal complement* of M) and the map

$$P_M(h) = m, \; h = m + m^{\perp}, \; m \in M, \; m^{\perp} \in M^{\perp}$$

is a projection such that $\|P_M\| \leqslant 1$.

Definition 2.3.4
Let $A \in \mathscr{B}(X, Y)$ and let $y^* \in Y^*$. Then the linear form x^* defined by $x^*(x) = y^*(Ax)$ clearly belongs to X^*. Putting $x^* = A^*y^*$, i.e.

$$\langle A^*y^*, x \rangle_{X^*, X} = \langle y^*, Ax \rangle_{Y^*, Y} \tag{2.18}$$

Then $A^* \in \mathscr{B}(Y^*, X^*)$ and $\|A^*\| = \|A\|$. A^* is called the *dual* of A. Note that if X and Y are Hilbert spaces, then we identify X^* with X and Y^* with Y and then $A^* \in \mathscr{B}(Y, X)$. However, in this case, the dual operator is usually defined by

$$\langle A^*h_1, h_2 \rangle_X = \langle h_1, Ah_2 \rangle_Y, \quad h_1 \in Y, h_2 \in X \tag{2.19}$$

and it is this definition of the dual (or *adjoint*) which we shall take in the Hilbert space case. It is easy to see that the adjoints defined by (2.18) and (2.19) (when X, Y are Hilbert spaces) are related by

$$A^*_{(2.19)} = j_X^{-1} A^*_{(2.18)} j_Y$$

We shall not preserve the distinction; the context should make clear which adjoint is intended.

The simplest bounded operators are, of course, matrices defined on a finite dimensional space (every such matrix defining a bounded operator); the Jordan decomposition gives a complete structure theory in this case. The next most simple class of operators are the compact operators $\mathscr{B}_0(X, Y) \subseteq \mathscr{B}(X, Y)$ defined as follows:

$$C \in \mathscr{B}_0(X, Y) \quad \text{iff } \overline{C(X_b)} \text{ is compact in } Y \tag{2.20}$$

for every bounded subset X_b of X. There is also a complete spectral theory for these operators, as we shall see later.

Example 2.3.5
An important example of a bounded operator on l^2 is the left-shift operator A_l defined by

$$A_l(x_1, x_2, \ldots)^T = (x_2, x_3, \ldots)^T, \quad \forall x = (x_1, x_2, \ldots)^T \in l^2 \tag{2.21}$$

The adjoint of A_l is the right-shift operator A_r given by

$$A_r x = A_l^* x = (0, x_1, x_2, \ldots)^T \tag{2.22}$$

Clearly, $\|A_r\| = \|A_l\| = 1$.

The remaining class of operators which we chall consider are those which are not bounded, but which have the following redeeming property

Definition 2.3.6
Let A be a (linear) operator from X into Y. If $x_n \in \mathscr{D}(A)$ and

$$(x_n \to x \text{ and } Tx_n \to v) \to (x \in \mathscr{D}(A) \text{ and } v = Tx) \tag{2.23}$$

we say that A is *closed*. A is *closable* if

$$(x_n \in \mathscr{D}(A), x_n \to 0 \text{ and } Ax_n \to v) \to v = 0 \tag{2.24}$$

Denote the class of closed operators by $\mathscr{C}(X,Y)$.

Theorem 2.3.7 (closed graph theorem)

$$A \in \mathscr{C}(X,Y) \text{ and } \mathscr{D}(A) = X \to A \in \mathscr{B}(X,Y). \quad \square$$

As ε ı example consider the operator $A = \mathrm{d}/\mathrm{d}x$ on $C([0,1])$ (*cf.* (2.2.9)(a)). Then A is unbounded and closed with domain $C^1([0,1])$. Suppose, however, that instead of the usual norm on $C([0,1])$ we impose the L^2 norm, then $\mathrm{d}/\mathrm{d}x$ is no longer even closable (in $C([0,1])$!). To define the closure of $\mathrm{d}/\mathrm{d}x$ in the L^2 norm, we must introduce the Sobolev spaces (which we do in Section 2.5).

The most important operators for us are the densely defined ones. We can then extend the notion of dual or adjoint as in (2.18). In fact, we have

$$\langle g,Au\rangle = \langle A^*g,u\rangle,\ u \in \mathscr{D}(A),\ g \in \mathscr{D}(A^*) \tag{2.25}$$

with $\overline{\mathscr{D}(A^*)} = Y^*$. If A is closable and densely defined, then A^{**} is the closure of A.

2.4 Spectral Theory

Definition 2.4.1
Let A be an operator with $\mathscr{D}(A), \mathscr{R}(A) \subseteq X$, where X is now a *complex* Banach space. For any complex number λ, the operator $(\lambda I - A)$ may or may not be one-to-one. For those λs for which $(\lambda I - A)^{-1} \in \mathscr{B}(X)$, we write

$$R(\lambda;A) = (\lambda I - A)^{-1} \qquad (\textit{resolvent} \text{ of } A) \tag{2.26}$$

and call the set of all such λs the *resolvent set* $\rho(A)$ of A. $\sigma(A) = \mathscr{C}\backslash\rho(A)$ is called the *spectrum* of A. Clearly, $\sigma(A)$ consists of three disjoint sets $\sigma_P(A)$, $\sigma_C(A)$, $\sigma_R(A)$ defined by

$$\begin{aligned}
\sigma_P(A) &= \{\lambda \in \sigma(A) : (\lambda I - A) \text{ is not } 1-1\} \\
\sigma_C(A) &= \{\lambda \in \sigma(A) : \overline{(\lambda I - A)^{-1}} = X\} \\
\sigma_R(A) &= \{\lambda \in \sigma(A) : \overline{(\lambda I - A)^{-1}} \neq X\}
\end{aligned}$$

$R(\lambda; A)$ is analytic on $\rho(A)$, which is the natural domain of analyticity of $R(\lambda;A)$, and we have the *resolvent equation*

$$R(\lambda; A) - R(\mu; A) = (\mu - \lambda)R(\lambda; A)R(\mu; A), \forall\, \lambda, \mu \in \rho(A) \tag{2.27}$$

The most important application of spectral theory is in the representation of operators in terms of simpler operators, as in the Jordan decomposition of matrices. In order to extend the latter theory to operators on Banach spaces, we must introduce the functional calculus due to Dunford and Taylor. Consider first the bounded case, i.e. $A \in \mathcal{B}(X)$.

Definition 2.4.2

Let $\mathcal{F}(A)$ denote the class of functions which are analytic on some neighbourhood of $\sigma(A)$. Now, if $f \in \mathcal{F}(A)$ and $U \supseteq \sigma(A)$ is an open set such that f is analytic on $U \cup \partial U$ where ∂U consists of a finite number of positively oriented rectifiable Jordan curves, then we define the value of the function f on the operator A by

$$f(A) = \frac{1}{2\pi i} \int_{\partial U} f(\lambda)R(\lambda; A)\mathrm{d}\lambda \tag{2.28}$$

Note that $f(A)$ depends only on f and A and not on U.

An important property of this definition is that if $f,g \in \mathcal{F}(A)$ then $f \cdot g \in \mathcal{F}(A)$ and $f(A) \cdot g(A) = (f \cdot g)(A)$. It follows from this that we have the *spectral mapping theorem*:

$$f(\sigma(A)) = \sigma(f(A)), \forall\, f \in \mathcal{F}(A) \tag{2.29}$$

Definition 2.4.3

A subset of $\sigma(A)$ is called a *spectral set* if it is open and closed in $\sigma(A)$.

Let $e \in \mathcal{F}(A)$ be such that

$$e(\lambda) = \begin{cases} 1 \text{ if } \lambda \in \sigma \\ 0 \text{ if } \lambda \in \sigma(A) \backslash \sigma \end{cases} \tag{2.30}$$

where σ is a spectral set. We then write $E(\sigma; A)$ (or $E(\sigma)$) $= e(A)$; clearly, $E(\sigma)$ is a projection operator. Put $X_\sigma = E(\sigma)X$, and $A\sigma = A|_{X_\sigma}$. Then,

$$AX_\sigma \subseteq X_\sigma \tag{2.31}$$

$$\sigma(A_\sigma) = \sigma \tag{2.32}$$

and

$$f(A)_\sigma = f(A_\sigma), \forall\, f \in \mathcal{F}(A) \tag{2.33}$$

If $\sigma(A)$ consists of a finite number $\sigma_1,\ldots,\sigma_n$ of spectral sets, then we have

$$E(\sigma_i)^2 = E(\sigma_i),\ E(\sigma_i)E(\sigma_j) = 0 \text{ if } i \neq j$$

and

$$I = \sum_{i=1}^{n} E(\sigma_i),\ X = \bigoplus_{i=1}^{n} X_{\sigma_i} \tag{2.34}$$

In finite-dimensional space, the spectral sets are just singletons consisting of the eigenvalues of the operator, and we can derive the Jordan decomposition from the above results.

Theorem 2.4.4
If $A \in \mathscr{B}_0(X)$, then the spectrum of A is countable with no accumulation point except possibly the origin. Every $\lambda \in \sigma(A)$ is a pole of A of some (varying) finite order ν and the space

$$E(\lambda)X = \{x : (A - \lambda I)^{\nu} x = 0\}$$

is finite dimensional.

Now let A' be a closed operator. Then we can use the above results by defining the homeomorphism $\Phi: \mathscr{C} \cup \{\infty\} \to \mathscr{C} \cup \{\infty\}$ by

$$\Phi(\lambda) = (\lambda - \alpha)^{-1},\ \lambda \neq \alpha$$

$$\Phi(\infty) = 0$$

$$\Phi(\alpha) = \infty$$

for any fixed $\alpha \in \rho(A')$. Then, if $A = -R(\alpha; A')$, we have

$$\Phi(\sigma(A') \cup \{\infty\}) = \sigma(A)$$

and if $\mathscr{F}'(A')$ denotes the set of functions analytic on a neighbourhood of $\sigma(A')$ and at infinity, then

$$f(\Phi^{-1}(.)) \in \mathscr{F}(A) \leftrightarrow f(.) \in \mathscr{F}'(A')$$

We can therefore define $f(A')$ by

$$f(A') = f(\Phi^{-1}(A)),\ f \in \mathscr{F}'(A')$$

and it follows that

$$f(A') = f(\infty)I + \frac{1}{2\pi i} \int_{\Gamma} f(\lambda) R(\lambda; A') d\lambda \tag{2.35}$$

where Γ consists of a finite number of Jordan arcs containing $\sigma(A')$. Note that $f(A')$ is independent of $\alpha \in \rho(A')$, and that we have the spectral mapping theorem

$$\sigma(f(A')) = f(\sigma(A') \cup \{\infty\}),\ f \in \mathscr{F}'(A') \tag{2.36}$$

It now follows easily from theorem 2.4.4 that we have the following corollary.

Corollary 2.4.5
If A is a closed self-adjoint operator on a Hilbert space H (i.e. $A = A^*$) and $R(\lambda; A)$ is compact (for some, and hence all, $\lambda \in \rho(A)$) then the spectrum of A consists of just eigenvalues λ_i with finite multiplicity such that $|\lambda_i| \to \infty$ as $i \to \infty$; moreover, there is a complete orthonormal set of eigenvectors ϕ_i such that $(\forall x \in H)$

$$x = \sum_{i=1}^{\infty} \langle x, \phi_i \rangle \phi_i$$

$$Ax = \sum_{i=1}^{\infty} \lambda_i \langle x, \phi_i \rangle \phi_i \tag{2.37}$$

$$R(\lambda; A)x = \sum_{i=1}^{\infty} \frac{1}{\lambda - \lambda_i} <x, \phi_i> \phi_i, \ \lambda \in \rho(A)$$

Note that $(\lambda_i)_{1 \leq i < \infty}$ is the sequence of eigenvalues counted according to multiplicity.

2.5 Distribution theory, weak topologies and Sobolev spaces

In this section we shall give a short introduction to distribution theory, which we require for the definition of the Sobolev spaces and Schwartz's kernel theorem (see also Horvath, 1966 and Treves, 1967).

Definition 2.5.1
A *topoligical vector space* E (over F) is a vector space over F with a topology such that the mappings

(a) $(x,y) \to x + y$ from E **x** E into E
(b) $(\lambda,x) \to \lambda x$ from F **x** E into E

are continuous.

Definition 2.5.2
A set $A \subseteq E$ is *absorbing* if $\forall x \in E, \exists \alpha > 0$ such that $x \in \lambda A, \forall \lambda \in F$ with $|\lambda| \geq \alpha$; $A \subseteq E$ is called *balanced* if $\lambda A \subseteq A$, $\forall \lambda \in F$ with $|\lambda| \leq 1$.

A *filter* $\mathcal{F}$ on a set X is a collection of subsets of X such that

(F1) If $A \subseteq X$ and $A \supseteq B \in \mathcal{F}$ then $A \in \mathcal{F}$

(F2) $A, B \in \mathcal{F} \to A \cap B \in \mathcal{F}$

(F3) $\phi \notin \mathcal{F}$

A subset $\mathcal{B}(\neq \phi) \subseteq \mathcal{F}$ is a *basis* of $\mathcal{F}$ if $A, B \in \mathcal{B} \to \exists C \subseteq A \cap B$ such that $C \in \mathcal{B}$. A basis for the filter of neighbourhoods of a point x in a topological vector space is called a *fundamental system of neighbourhoods of* x. It is clear from the fact that the map $y \to y + x$ is a homeomorphism that the topology of a topological vector space is defined entirely by a fundamental system of neighbourhoods of 0.

Definition 2.5.3
A topological vector space is *locally convex* if each point has a fundamental system of convex neighbourhoods. Then we have the following theorem.

Theorem 2.5.4
Let E be a vector space and $\mathcal{G}$ be a collection of absorbing, balanced and convex subsets of E. If

$$\mathcal{N} = \{\bigcap_{i=1}^{n} (\lambda_i V_i) : \lambda_i > 0,\ V_i \in \mathcal{G}\}$$

then $\mathcal{N}$ is a fundamental system of neighbourhoods of zero of a unique locally convex topology on E. Moreover, $\mathcal{N}$ is equivalent (i.e., generates the same topology) to the system

$$\mathcal{B} = \{\lambda V : \lambda > 0,\ V = \bigcap_{i=1}^{n} U_i \text{ for some } U_i \in \mathcal{G}\}. \quad \square \tag{2.38}$$

It is important that locally convex spaces can be defined in terms of seminorms (i.e. maps $q: E \to \mathcal{R}_+$ such that $q(\lambda x) = |\lambda| q(x), \forall \lambda \in F, x \in E$, and $q(x + y) \leqslant q(x) + q(y), \forall x, y \in E$). In fact, if q is a seminorm, then the set

$$V = \{x : q(x) \leqslant 1\}$$

is balanced, absorbing and convex and so by (2.38), if $\{q_i\}_{i \in I}$ is a family of seminorms, the sets

$$V_{i_1,\ldots,i_n,\epsilon} = \{x : q_{i_k}(x) \leqslant \epsilon \text{ for } 1 \leqslant k \leqslant n\},\ i_k \in I$$

form a fundamental system of neighbourhoods of zero of a locally convex space, and conversely, any locally convex space can be defined by *all* continuous seminorms

It follows that if E and F are locally convex spaces defined by families of seminorms $\{p_i\}_{i \in I}$, $\{q_j\}_{j \in J}$, respectively, then a linear map $f: E \to F$ is continuous iff for any q_j, $\exists p_i$ and $M > 0$ such that

$$q_j(x) \leqslant M p_i(x),\ \forall x \in E \tag{2.39}$$

Examples 2.5.5
We shall now define the spaces which are of the greatest importance in distribution theory.

(a) Let Ω be an open set in $\mathcal{R}^n$. We have already met the space $C^k(\bar{\Omega})$ and seen that it is a Banach space. However, in distribution theory we wish to generalise the notions of derivative, Fourier transform etc., by transposition in some duality. This latter process corresponds, when suitably restricted, to integration by parts. Since we do not want to have the 'boundary term' in this expression we consider spaces whose members vanish outside compact subsets of Ω. Hence, to begin, let $K \subseteq \Omega$ be compact, and let $\mathscr{D}(K)$ be the space of functions defined in (a neighbourhood of) K whose derivatives of all orders exist and are continuous, and which vanish outside K. Topologise $\mathscr{D}(K)$ as a locally convex space with the seminorms

$$q_{k,K}(f) = \max_{x \in K} |(\partial/\partial x)^p f(x)|, \forall p \in N^n \tag{2.40}$$

Now let $\mathscr{D}(\Omega) = \cup \mathscr{D}(K)$ (union over all compact subsets of Ω) and give $\mathscr{D}(\Omega)$ the finest locally convex topology for which all the inclusion maps $\mathscr{D}(K) \to \mathscr{D}(\Omega)$ are continuous.

(b) $\mathscr{E}(\Omega)$ denotes the space of functions which are infinitely differentiable (with no restriction on their supports), with the family of seminorms (2.40) for all $p \in N^n$ and all compact $K \subseteq \Omega$.

(c) A space which is used in Fourier transform theory of distributions is

$$\mathscr{S} = \{f \in \mathscr{E}(\mathscr{R}^n) : \forall p \in N^n,\ k \in Z,\ \varepsilon > 0,\ \exists \rho > 0$$

with

$$|(1 + |x|^2)^k (\partial/\partial x)^p f(x)| \leqslant \varepsilon \quad \text{for } |x| > \rho\} \tag{2.41}$$

The space $\mathscr{S}$ is a locally convex space with the system of seminorms

$$q_{k,p}(f) = \max_{x \in R^n} |(1 + |x|^2)^k (\partial/\partial x)^p f(x)| \tag{2.42}$$

Note that $\mathscr{D}(\Omega) \subseteq \mathscr{E}(\Omega)$, $\mathscr{D}(\mathscr{R}^n) \subseteq \mathscr{S} \subseteq \mathscr{E}(\mathscr{R}^n)$, each inclusion being dense and continuous.

Definition 2.5.6

Let E be a topological vector space and E^* the dual space. The duality $\langle \cdot, \cdot \rangle_{E,E^*}$ defines two systems of seminorms:

$$q^{y^*}: x \to |\langle x, y^* \rangle| \text{ is a seminorm on } E \text{ for each } y^* \in E^*$$

$$q_x: y^* \to |\langle x, y^* \rangle| \text{ is a seminorm on } E^* \text{ for each } x \in E$$

The systems $\{q^{y^*}; y^* \in E^*\}$, $\{q_x; x \in E\}$ define locally convex topologies on E and E^* called the weak and weak* topologies, respectively (the notations $\sigma(E, E^*)$, $\sigma(E^*, E)$ are used for these topologies).

It is clear that a sequence $\{x_n\} \subseteq E$ (or $\{y_n^*\} \subseteq E^*$) converges weakly to x (or weakly* to y^*) iff

$$\lim_n \langle x_n, x^* \rangle = \langle x, x^* \rangle, \forall x^* \in E^* \tag{2.43}$$

$$(\text{or } \lim_n \langle y, y_n^* \rangle = \langle y, y^* \rangle, \forall y \in E)$$

Theorem 2.5.7
The unit sphere of a Hilbert space H is compact in the weak topology (it is, of course, not compact in the norm topology, unless H is finite dimensional). Also, a closed, convex subset of H is weakly closed. □

Definition 2.5.8
The dual space $\mathscr{D}'(\Omega)$ of $\mathscr{D}(\Omega)$ is called the *space of distributions* on Ω. $\mathscr{S}'$ is the space of *tempered distributions* and $\mathscr{E}'(\Omega)$ the *distributions with compact support* in Ω. Clearly,

$$\mathscr{E}'(\Omega) \subseteq \mathscr{D}'(\Omega), \mathscr{S}' \subseteq \mathscr{D}'(\mathscr{R}^n)$$

Definition 2.5.9
The main reason for defining distributions is to give a rigorous meaning to the δ 'function' and to generalise differentiation, Fourier transform etc., to such objects; this generalisation is done by transposition in the $\mathscr{D}'(\Omega)$, $\mathscr{D}(\Omega)$ duality. For example:

(a) Let $T \in \mathscr{D}'(\Omega)$ and $p \in N^n$ with $|p| = p_1 + \ldots + p_n$. Then we define the *derivative* $(\partial/\partial x)^p$ of T to be the distribution given by

$$\langle(\partial/\partial x)^p T, \phi\rangle = (-1)^{|p|}\langle T, (\partial/\partial x)^p \phi\rangle, \forall \phi \in \mathscr{D}(\Omega) \tag{2.44}$$

(b) If $T \in \mathscr{S}'$, we define the *Fourier transform* $\mathscr{F}T$ of T by

$$\langle \mathscr{F}T, \phi\rangle = \langle T, \mathscr{F}\phi\rangle, \forall \phi \in \mathscr{S} \tag{2.45}$$

Examples 2.5.10
(a) The distribution δ is defined by

$$\langle \delta, \phi\rangle = \phi_0$$

Clearly,

$$\langle(\partial/\partial x)^p \delta, \phi\rangle = (\ 1)^{|p|}((\partial/\partial x)^p \phi)(0)$$

(b) A function frequently used in control engineering is the step function, defined by

$$U(x) = \begin{cases} 1 \text{ if } x \geqslant 0 \\ 0 \text{ if } x < 0 \end{cases}$$

It is easy to see that the map $\bar{U}: \mathscr{D}(\mathscr{R}) \to F$ given by

$$\bar{U}(\phi) = \int_{-\infty}^{\infty} U(x)\phi(x)\mathrm{d}x = \int_0^{\infty} \phi(x)\mathrm{d}x$$

belongs to $\mathscr{D}'(\mathscr{R})$. Then,

$$\langle(\partial/\partial x)\bar{U}, \phi\rangle = -\langle \bar{U}, \partial\phi/\partial x\rangle = \langle \delta, \phi\rangle$$

Hence,

$$(\partial/\partial x)\bar{U} = \delta$$

(c) More generally, if $f{:}\Omega \to F$ is any locally integrable function, then the associated map $\bar{f}{:}\ \mathscr{D}(\Omega) \to F$ given by

$$\bar{f}(\phi) = \langle \bar{f}, \phi \rangle = \int_\Omega f(x)\, \phi\, (x)\mathrm{d}x \tag{2.46}$$

is a distribution. In particular, if $1(x) = 1, \forall x \in \mathscr{R}^n$, then

$$\langle \bar{1}, \phi \rangle = \int_{\mathscr{R}^n} \phi(x)\mathrm{d}x$$

(d)

$$\langle \mathscr{F}\delta, \phi \rangle = \langle \delta, \mathscr{F}\phi \rangle = \int_{\mathscr{R}^n} \phi(x)e^{-2\pi i\langle x,\xi\rangle}\mathrm{d}x|_{\xi=0}$$
$$\int_{\mathscr{R}^n} \phi(x)\mathrm{d}x = \langle \bar{1}, \phi \rangle$$

Hence, $\mathscr{F}\delta = \bar{1}$.

Theorem 2.5.11 (Schwartz's Kernel theorem)
Let $X \subseteq \mathscr{R}^m$, $Y \subseteq \mathscr{R}^n$ be open sets. Then $X \times Y \subseteq \mathscr{R}^{m+n}$ is open and we have

$$\mathscr{D}'(X \times Y) \cong \mathscr{L}(\mathscr{D}(Y); \mathscr{D}'(X))$$

i.e. the space of distributions on $X \times Y$ is isomorphic (as a topological vector space) to the space of continuous linear maps from $\mathscr{D}(Y)$ into $\mathscr{D}'(X)$. □

This theorem means that if we are given a continuous linear map $L{:}\mathscr{D}(Y) \to \mathscr{D}'(X)$ we can associate with it a distribution $K \in \mathscr{D}'(X \times Y)$, and it can be shown that if $\phi \in \mathscr{D}(X)$, $\psi \in \mathscr{D}(Y)$, then we have

$$\langle K, \phi\psi \rangle = \langle L\, \psi, \phi \rangle \tag{2.47}$$

where $\phi\psi{:}(x, y) \in \mathscr{R}^{m+n} \to \phi(x)\ \psi\ (y)$. Sometimes one writes (somewhat incorrectly)

$$(L\psi)(x) = \int K(x, y)\psi(y)\mathrm{d}y$$

If we regard $L^2(X \times Y)$ as a subset of $\mathscr{D}'(X \times Y)$, then it can be shown that $L^2(X \times Y) \cong \mathscr{B}_0^n(L^2(X), L^2(Y))$, where the latter space is the subspace of the space of compact operators N from $L^2(X)$ into $L^2(Y)$ which are *nuclear*, i.e. which can be written in the form

$$Nf = \sum_k \lambda_k \langle f, f_k \rangle g_k$$

where $f_k(g_k)$ is a basis of $L^2(X)$ (respectively $L^2(Y)$) and $\Sigma|\lambda_k| < \infty$; then we can write

$$(Nf)(y) = \int_X N(x, y)f(x)\mathrm{d}x$$

where $f \in L^2(X), N(x, y) \in L^2(X \times Y)$.

We come now to introduce the Sobolev spaces which are indispensible in the theory of partial differential equations. Note first that, by (2.46),

$$\mathscr{D}(\Omega) \subseteq L^p(\Omega) \subseteq \mathscr{D}'(\Omega),\ 1 \leqslant p \leqslant \infty$$

(since any locally L^p function is locally L^1). Hence it makes sense to say that a distribution is in L^p.

Definition 2.5.12
We define the space

$$H^{p,m}(\Omega) = \{f \in \mathscr{D}'(\Omega) : (\partial/\partial x)^\alpha f \in L^p(\Omega),\ |\alpha| \leqslant m\}$$

with the norm

$$\|f\|_{p,m} = \left\{ \sum_{|\alpha| \leqslant m} \int_\Omega |(\partial/\partial x)^\alpha f(x)|^p \mathrm{d}x \right\}^{1/p} \tag{2.48}$$

It is easy to see that $H^{p,m}(\Omega)$ is a Banach space and that $H^{2,m}(\Omega)$ is a Hilbert space under the inner product

$$\langle f, g \rangle_{p,m} = \sum_{|\alpha| \leqslant m} \int_\Omega (\partial/\partial x)^\alpha f(x) \overline{(\partial/\partial x)^\alpha g(x)} \mathrm{d}x \tag{2.49}$$

Definition 2.5.13
Denote by $H_0^{p,m}(\Omega)$ $(1 \leqslant p \leqslant \infty, m \geqslant 1)$ the closure of $\mathscr{D}(\Omega)$ in $H^{p,m}(\Omega)$. Put

$$H^{p',-m}(\Omega) = H_0^{p,m}(\Omega))^*,\ p' = p/(p-1),\ 1 \leqslant p < \infty$$

Then it can be shown that

$$H^{p',-m}(\Omega) = \{f \in \mathscr{D}'(\Omega): f = \sum_{|\alpha| \leqslant m} (\partial/\partial x)^\alpha g_\alpha,\ g_\alpha \in L^p(\Omega)\}$$

Note that if $\Omega = \mathscr{R}^n$, then $H_0^{p,m}(\mathscr{R}^n) = H^{p,m}(\mathscr{R}^n)$. In this case it is convenient to define the spaces H^s for arbitrary real s.

Definition 2.5.14
We define, for $s \in \mathscr{R}$

$$H^s = \{f \in \mathscr{S}'(\mathscr{R}^n): (1 + |\xi|^2)^{s/2} \hat{f} \in L^2(\mathscr{R}^n_\xi)\}$$

where $\hat{f}$ denotes the Fourier transform of f. For example, it is clear that $\delta \in H^{-n/2-\epsilon}$ for any $\epsilon > 0$.

Finally, in this section we should mention the Sobolev imbedding theorem, which takes several forms, one of which is as follows (Adams, 1975).

Theorem 2.5.15
Suppose that $\Omega \subseteq \mathscr{R}^n$ is an open bounded set with sufficiently smooth boundary. Then:

(a) if $m < k - n/p$, we have $H^{k,p}(\Omega) \subseteq C^m(\bar{\Omega})$;
(b) if $1/q > 1/p - k/n$, with $1 \leqslant p \leqslant \infty$, $1 \leqslant q \leqslant \infty$, $k \geqslant 1$, then we have the

imbedding

$$H^{k,p}(\Omega) \subseteq L^q(\Omega)$$

which is compact if $p,q < \infty$. □

2.6 Partial Differential Operators

A linear partial differential operator of order m is an expression

$$L = \sum_{|\alpha| \leq m} a_\alpha(x) D^\alpha = \sum_{k=0}^{m} \{ \sum_{\alpha_1 + \ldots + \alpha_n = k} a_{\alpha_1 \ldots \alpha_n}(x) D_1^{\alpha_1} \ldots D_n^{\alpha_n} \} \tag{2.50}$$

where $D_i = \partial/\partial x_i$, $\alpha \in N^n$ and $a_\alpha(x)$ is defined in a bounded open set $\Omega \subseteq R^n$. The highest-order terms (or *principal part*)

$$\sum_{|\alpha|=m} a_\alpha(x) D^\alpha$$

in a sense dominate the behaviour of this operator, and we associate with it a polynomial

$$p_m(\xi, x_0) \neq \sum_{|\alpha|=m} a_\alpha(x_0)\xi^\alpha, \text{ for each } x_0 \in \Omega$$

in $\mathscr{C}[\xi]$, $\xi \in \mathscr{C}^n$. If

$$p_m(\xi, x_0) \neq 0$$

for each $0 \neq \xi \in R^n$ we say that (2.50) is *elliptic* (at x_0). If the coefficients a_α are real, then m must be even. Put $m = 2q$. The operator (2.50) is *(uniformly) strongly elliptic* in Ω if

$$(-1)^q \mathrm{Re}\, p_{2q}(\xi, x) \geq c|\xi|^{2q}, \forall x \in \Omega \tag{2.51}$$

for some constant $c > 0$.

If the coefficients a_α are sufficiently differentiable, we can write the operator (2.50) (with $m = 2q$) in the 'divergence form'

$$Lu = \sum_{0 \leq |\beta|, |\gamma| \leq q} (-1)^{|\beta|} D^\beta (a^{\beta\gamma}(x) D^\gamma u) \tag{2.52}$$

and so, integrating by parts, we have

$$\langle v, Lu \rangle_{L^2} = B(v, u) \triangleq \sum_{0 \leq |\beta|, |\gamma| \leq q} \langle D^\beta v, a^{\beta\gamma} D^\gamma u \rangle_{L^2} \tag{2.53}$$

for all $u, v \in \mathscr{D}(\Omega)$. Note that B is a bilinear form.

In order to prove the existence of solutions to elliptic partial differential equations, we use the Lax-Milgram theorem in conjunction with the following inequality, due to Garding.

Theorem 2.6.1

If L is strongly elliptic in Ω, the coefficients $a^{\beta\gamma}$ are bounded in Ω and

$$|a^{\beta\gamma}(x) - a^{\beta\gamma}(y)| \leq f(\|x - y\|),\ x,y \in \Omega,\ |\beta| = |\gamma| = q$$

where $f(t) \to 0^+$ as $t \to 0^+$, then we have

$$\text{Re } B(u, u) \geqslant c_1\|u\|_{2,m}^2 - c_2\|u\|_{2,0}^2, \forall u \in H_0^{2,m}(\Omega)$$

for some constants $c_1 > 0$, and c_2. □

The *Dirichlet problem* consists of determining a solution of the problem

$$\begin{cases} Lu = f \text{ in } \Omega \\ \dfrac{\partial^j u}{\partial \nu^j} = g_j \text{ on } \partial\Omega, \ 0 \leqslant j \leqslant m - 1 \end{cases}$$

for some functions f, g_j where ν is the (outward) normal to Ω. If $g_j \in C^{2m}(\partial\Omega)$, it can be shown that there is no loss of generality in taking each $g_j \equiv 0$. Thus, our problem becomes

$$\begin{cases} Lu = f, f \in L^2(\Omega) \\ \dfrac{\partial^j u}{\partial \nu^j} = 0 \end{cases} \tag{2.54}$$

By a *solution* of (2.54) we shall mean an element $u \in H_0^{2,m}(\Omega)$ such that

$$B(\phi, u) = \langle \phi, f \rangle_{L^2}, \forall \phi \in \mathscr{D}(\Omega) \tag{2.55}$$

This makes sense in view of (2.53).

Theorem 2.6.2

Suppose that L satisfies the conditions of theorem 2.6.1; then there exists a constant c_2 such that for all $c \geqslant c_2$, the Dirichlet problem for the operator $L + c$ has a unique solution for any $f \in L^2(\Omega)$.

Proof: Consider the bilinear form $B_1(u, v) = B(u, v) + c\langle u, v \rangle_{L^2(\Omega)}$, for all $u,v \in H_0^{2,m}(\Omega)$. Since L is associated with B, we have, by Garding's inequality.

$$\text{Re } B(u,u) \geqslant c_1\|u\|_{2,m}^2 - c_2\|u\|_{L^2(\Omega)}^2$$

for some constants $c_1 > 0$, c_2. Hence,

$$\text{Re } B_1(u,u) > c_1\|u\|_{2,m}^2$$

if $c \geqslant c_2$; moreover, B_1 is associated with $L + c$. Now, the linear form

$$\phi \to \langle \phi, f \rangle_{L^2}, \ \phi \in H_0^{2,m}(\Omega)$$

is continuous on $H_0^{2,m}(\Omega)$ and so by proposition 2.2.10, there exists $u \in H_0^{2,m}(\Omega)$ such that (2.55) holds, for each $\phi \in H_0^{2,m}(\Omega)$. □

Let us denote, for simplicity, $H_0^m(\Omega) \triangleq H_0^{2,m}(\Omega)$, $H^m(\Omega) \triangleq H^{2,m}(\Omega)$, and define the operator A with domain $\mathscr{D}(A) = H^{2m}(\Omega) \cap H_0^m(\Omega)$ by

$$(Au)(x) = Lu(x), \ u \in \mathscr{D}(A) \tag{2.56}$$

where L is given by (2.50). Then, using similar methods to those above we

Theorem 2.6.3
Let Ω be a bounded domain in $\mathscr{R}^n$ with sufficiently smooth boundary and let L be strongly elliptic in Ω, with coefficients $a_\alpha \in C^j(\bar{\Omega})$, $j = \max(0, |\alpha| - m)$. Then the operator A given in (2.56) is a closed operator defined in $L^2(\Omega)$ with domain $\mathscr{D}(A)$. Moreover, the resolvent $(\lambda I - A)^{-1}: L^2(\Omega) \to L^2(\Omega)$ exists for all $\lambda \in \mathscr{C}$ such that λ belongs to the sector

$$\{\lambda: \tfrac{1}{2}\pi < \arg(\lambda + k) < (3/2)\pi, \text{ some } k > 0\}$$

and

$$\|(\lambda I - A)^{-1}\|_{\mathscr{L}(L^2(\Omega))} \leqslant \frac{C}{|\lambda| + 1} \tag{2.57}$$

for some $C > 0$. □

For more details on partial differential equations see Friedman (1969), Agmon (1965), Dunford and Schwartz (1963) and Lions and Magenes (1972).

2.7 Generalisations

We shall need, in addition to the usual $L^p(\Omega)$ spaces introduced above, the more general spaces $L^p(I; X)$, where $I \subseteq \mathscr{R}$, of strongly measurable functions $f: I \to X$ with values in a Banach space X such that

$$\|f\| = \left(\int_I \|f(t)\|_X^p \mathrm{d}t\right)^{1/p} < \infty, \; p < \infty$$

$$\|f\| = \operatorname*{ess\,sup}_{t \in I} \|f\|_X, \; p = \infty$$

These spaces can be regarded as subspaces of spaces of vector-valued distributions

$$\mathscr{D}'(I; X) = \mathscr{L}(\mathscr{D}(I); X)$$

(assuming I is open) in the obvious way.

Theorem 2.7.1 (Young's inequality)
If $f \in L^p(\mathscr{R}^n)$, $g \in L^q(\mathscr{R}^n)$ then $f^*g \in L^r(\mathscr{R}^n)$, where $1 \leqslant p, q, r \leqslant \infty$ and $1/r = 1/p + 1/q - 1$. Moreover,

$$\|f^*g\|_{L^r} \leqslant \|f\|_{L^p} \cdot \|g\|_{L^q}. \quad \square$$

(Here, $(f^*g)(x) = \int_{\mathscr{R}^n} f(x - y)g(y)\mathrm{d}y$.) This can easily be generalised as follows.

Corollary 2.7.2
If $B(t)$, $t \geqslant 0$, is a bounded operator-valued function belonging to $L^p(]0, \infty[; (X))$ and $x(t) \in L^q(]0, t[; X)$, then

$$\|B^*x\|_{L^r(]0,T[;X)} \leqslant \|B\|_{L^p(]0,\infty[;(X))}\|x\|_{L^q(]0,T[;X)}$$

where

$$(B^*x)(t) = \int_0^t B(t-s)x(s)\mathrm{d}s. \quad \square$$

2.8 The Fréchet Derivative

Let X, Y be Banach spaces and let $f{:}X \to Y$ be a (nonlinear) map from X into Y. Then we say that f is *Fréchet differentiable* at $x \in X$ if there exists a linear operator $L(x) \in \mathscr{L}(X, Y)$ such that

$$\frac{\|f(x+h) - f(x) - L(x)y\|_Y}{\|h\|_X} \to 0$$

as $h \to 0$ in X. We shall write

$$f'(x)h = L(x)h$$

Chapter 3
Semigroup theory, evolution equations and system examples

3.1 Introduction

The state-space theory of the control of lumped pararameter systems is based on the solution of the differential equation

$$\dot{x} = Ax, x(0) = x_0 \in \mathscr{R}^n \tag{3.1}$$

defined on n-dimensional Euclidean space $\mathscr{R}^n$ (where A is a matrix) and that of the perturbed form

$$\dot{x} = Ax + Bu \tag{3.2}$$

These solutions are given, respectively, by

$$x(t) = e^{At}x_0 \tag{3.3}$$

and

$$x(t) = e^{At}x_0 + \int_0^t e^{A(t-s)}Bu(s)ds \tag{3.4}$$

(the variation of constants formula).

In the case of distributed parameter systems, the mathematical description is usually given by a partial differential, delay or integro-differential equation. Because of the simplicity of dealing with the finite dimensional systems given by equations of the form (3.1) or (3.2), we would like to be able to write the equation of a distributed system formally in the same way. In order that (3.1) represents such a system, the state x will belong in general to some Banach space X, or, more usually, to a Hilbert space H. However, it turns out that the operator A is no longer bounded and so we cannot define the operator e^{At} (at least in the case of a differential operator A; later we shall consider a bounded approximation to a delay equation).

When A is an unbounded operator it is therefore not obvious what the solution should be. However, we can obtain insight into this question by considering the properties which the solution of any reasonable physical system ought to possess. By comparison with the finite-dimensional system these are

(*a*) $x(0;x_0) = x_0$
(*b*) $x(t_1 + t_2;x_0) = x(t_1;x(t_2;x_0)) = x(t_2;x(t_1;x_0))$ (3.5)
(*c*) $x(t;x_0)$ is continuous in t and x_0 (in some sense) where $x(t;x_0)$ is the (assumed unique) solution of (3.1) through x_0 at $t = 0$.

The physical interpretation of (*a*) is clear; that of (*b*) is that the state at which we arrive by following the trajectory from x_0 for a time $t_1 + t_2$ is the same as the state reached by going for a time t_1 (or t_2) followed by a time t_2 (or t_1). Condition (*c*) is essentially a type of local stability condition which says that if we change t or x_0 slightly, then the solution should not change too much. It is these properties which lead us to define the notion of (semi-) group of operators, to be introduced in the next section. Some systems (for example, the heat equation) have solutions which become smoother (in a sense to be discussed below) and so it is not possible to define the solution (at least on the same Hilbert space) backwards in time; for this reason we consider distributed systems generally on the time interval $[0,\infty)$. A notable exception is the wave equation.

3.2 Linear semigroup theory

Abstracting the conditions (3.5(*a*), (*b*), (*c*)) we introduce the following definition.

Definition 3.2.1
A (*strongly continuous*) *semigroup of operators* is an operator-valued function $T:\mathscr{R}_+ \to \mathscr{B}(X)$ such that

(*a*) $T(0) = I$
(*b*) $T(t_1 + t_2) = T(t_1)T(t_2)$ (3.6)
(*c*) $\lim_{t \to 0^+} T(t)x = x, \forall x \in X$ (i.e. T is strongly continuous at $t = 0$)

Returning to the case of a finite dimensional system (3.1), we have,

$$\lim_{t\to 0^+}\left\{\frac{e^{At}x - x}{t}\right\} = Ax,\ x \in \mathscr{R}^n$$

It is clear, therefore, that for infinite dimensional systems where we are expecting the solution through x_0 to be given by $T(t)x_0$, we should have

$$\lim_{t\to 0^+}\left\{\frac{T(t)x - x}{t}\right\} = Ax, x \in \mathscr{D}(A) \subseteq X \tag{3.7}$$

Note that, since A will generally be an unbounded operator, (3.7) will only hold on $\mathscr{D}(A)$.

Definition 3.2.2
A closed operator A with dense domain in X is called the (*infinitesimal*) *generator* of the semigroup T if (3.7) holds.

Theorem 3.2.3
Let $T(t)$ be a strongly continuous semigroup defined on X. Then:

(*a*) $\|T(t)\|$ is locally bounded on $[0,\infty)$;
(*b*) $T(t)$ is strongly continuous for all $x \in X$;
(*c*) $\omega_0 \overset{\Delta}{=} \inf\limits_{t>0} \dfrac{1}{t} (\log \|T(t)\|) = \lim\limits_{t\to\infty} \dfrac{1}{t} (\log \|T(t)\|) < \infty$;
(*d*) $\forall \omega > \omega_0$, $\exists M(= M(\omega))$ such that $\|T(t)\| \leqslant Me^{\omega t}$, $t \geqslant 0$.

Proof: (*a*) First, $\|T(t)\|$ is bounded on $[0,\delta]$ for some $\delta > 0$. For, if not, then there exists a sequence $\{t_n\}$ with $t_n \to 0^+$ such that $\|T(t_n)\| \geqslant n$. By a simple corollary of theorem 2.2.11, it follows that $\{\|T(t_n)x\|\}$ must be unbounded for some $x \in X$, which contradicts the strong continuity of $T(t)$ at the origin. Hence,

$$\|T(t)\| \leqslant M \text{ (say) for } t \in [0,\delta]$$

Now put $t = m\delta + \tau$ with $0 \leqslant \tau < \delta$. Then,

$$\|T(t)\| \leqslant \|T(\delta)\|^m \|T(\tau)\| \leqslant M^{1+m} \leqslant MM^{t/\delta} = Me^{\omega t}$$

if we set $\omega = (\log M)/\delta$.
(*b*) Let $t > 0$ and $|s| \leqslant t$. Then

$$\|T(t+s)x - T(t)x\| \leqslant \max\{\|T(t)\|, \|T(t+s)\|\}\cdot\|x - T(|s|)x\|$$

and (*b*) follows from the strong continuity at $t = 0$, by letting $s \to 0$.
(*c*) By definition of ω_0, if $\omega > \omega_0$, then $\exists\, t_0 > 0$ such that $t_0^{-1} \log \|T(t_0)\| < \omega$. Then, for every $t \geqslant t_0$, $\exists$ an integer n such that $nt_0 \leqslant t < (n+1)t_0$. Thus,

$$\begin{aligned} t^{-1} \log \|T(t)\| &\leqslant t^{-1} \log \|(T(t_0))^n T(t - nt_0)\| \\ &\leqslant t^{-1} n \log \|T(t_0)\| + t^{-1} \log M', \text{ some } M' > 0 \\ &\leqslant t_0^{-1} \log \|T(t_0)\| + t^{-1} \log M' \\ &< \omega + t^{-1} \log M' \end{aligned}$$

Since M' is fixed,

$$\limsup_{t\to\infty} \frac{\log \|T(t)\|}{t} \leqslant \omega_0$$

because $\omega > \omega_0$ is arbitrary. But

$$\omega_0 = \inf (t^{-1} \log \|T(t)\|) \leqslant \liminf (t^{-1} \log \| T(t)\|)$$

and so (*c*) is true.

(d) Let ω and t_0 be as in (c). Then, as above, $\exists t'_0$ such that

$$\|T(t)\| \leqslant e^{\omega t} \text{ for } t \geqslant t'_0$$

Now $\|T(t)\|$ is locally bounded, so $\|T(t)\| \leqslant M_0$ (say) on $[0,t'_0]$. Put $M'_0 = \max \{1,M_0\}$ and

$$M = \begin{cases} M'_0 \text{ if } \omega \geqslant 0 \\ M'_0 e^{-\omega t'_0} \text{if } \omega < 0 \end{cases}$$

Then

$$\|T(t)\| \leqslant M e^{\omega t}. \quad \square$$

We next want to relate the semigroup to the differential equation $\dot{x} = Ax$, defined by its infinitesimal generator. This is done in the following theorem.

Theorem 3.2.4
If $T(t)$ is a strongly continuous semigroup of operators on X, with generator A, then

(a) $T(t):\mathcal{D}(A) \to \mathcal{D}(A),\ t \geqslant 0$

(b) $\dfrac{d^n}{dt^n}(T(t)x) = A^n T(t)x = T(t)A^n x,\ x \in \mathcal{D}(A^n),\ t > 0$

(c) A is closed with $\overline{\mathcal{D}(A)} = X$

Proof: If $t \geqslant 0$ and $s > 0$, then

$$\frac{T(t+s)x - T(t)x}{s} = T(t)\frac{(T(s) - I)}{s}x = \frac{(T(s) - I)}{s}T(t)x \tag{3.8}$$

If $x \in \mathcal{D}(A)$, then the middle term has limit $T(t)Ax$ as $s \to 0$, by the strong continuity of $T(t)$. Hence the limits of the other terms exist and so by the third term, $T(t)x \in \mathcal{D}(A)$ and we obtain (a).

We prove (b) for $n = 1$ by noting that if $t > 0$ and $0 < s \leqslant t$, then

$$-\left\{\frac{T(t-s)x - T(t)x}{s}\right\} = T(t-s)\frac{(T(s) - I)x}{s}$$

The limit of this as $s \to 0$ together with (3.8) gives (b) for $n = 1$. The general case follows by induction.

To prove (*c*) let x be any element of X. Then, if $t > 0$, $s > 0$,

$$\frac{(T(s) - I)}{s} \int_0^t T(\tau)x\mathrm{d}\tau = \frac{1}{s} \int_0^t T(s + \tau)xd\tau - \frac{1}{s} \int_0^t T(\tau)xd\tau$$

$$= \frac{1}{s} \int_s^{t+s} T(\tau)x\mathrm{d}\tau - \frac{1}{s} \int_0^t T(\tau)x\mathrm{d}\tau$$

$$= \frac{1}{s} \int_0^s T(\tau)(T(t) - I)x\mathrm{d}\tau$$

$$\rightarrow (T(t) - I)x$$

as $s \rightarrow 0$, since the integrand is continuous. Hence,

$$\int_0^t T(\tau)x\mathrm{d}\tau \in \mathscr{D}(A)$$

However, again by the strong continuity of T,

$$x_n \triangleq \frac{1}{t_n} \int_0^{t_n} T(\tau)x\mathrm{d}\tau \rightarrow x$$

where t_n is a sequence tending to zero. Since $x_n \in \mathscr{D}(A)$ and x was arbitrary in X, $\overline{\mathscr{D}(A)} = X$.

To show that A is closed, note that, if $x \in \mathscr{D}(A)$ and $x^* \in X^*$, then

$$\langle x^*, T(t)x - x \rangle = \int_0^t \frac{\mathrm{d}}{\mathrm{d}\tau} \langle x^*, T(\tau)x \rangle \mathrm{d}\tau$$

$$= \int_0^t \langle x^*, T(\tau)Ax \rangle \mathrm{d}\tau$$

$$= \langle x^*, \int_0^t T(\tau)Ax\mathrm{d}\tau \rangle$$

By the Hahn-Banach theorem,

$$T(t)x - x = \int_0^t T(\tau)Ax\mathrm{d}\tau \tag{3.9}$$

Now, if $\{x_n\} \subseteq \mathscr{D}(A)$ is such that $x_n \rightarrow x$ and $Ax_n \rightarrow y$, then

$$\|T(s)Ax_n - T(s)y\| < Me^{\omega s} \|Ax_n - y\|$$

and so $T(s)Ax_n \rightarrow T(s)y$ uniformly on $[0,t]$. By (3.9)

$$T(t)x_n - x_n = \int_0^t T(\tau)Ax_n\mathrm{d}\tau$$

and by the Lebesgue dominated convergence theorem

$$T(t)x - x = \int_0^t T(\tau)y\,d\tau$$

Hence, $x \in \mathscr{D}(A)$ and $Ax = y$. □

For finite dimensional systems, we have $\mathscr{L}(e^{At}) = (sI - A)^{-1}$. It is important that this can be generalised to semigroups as the following result shows.

Proposition 3.2.5
If $T(t)$ is a strongly continuous semigroup with generator A, then $\text{Re}(s) > \omega \rightarrow s \in \rho(A)$ (where $\|T(t)\| \leqslant Me^{\omega t}$) and

$$R(s;A)x = \int_0^\infty e^{-st}T(t)x\,dt, \forall x \in X$$

Proof: Note first that the integral is well defined (in the Bochner sense), since the integrand is continuous and

$$\|e^{-st}T(t)x\| \leqslant Me^{(\omega-\sigma)t}\|x\|, \sigma = \text{Re}(s)$$

If $I(s)x$ denotes the integral then $\|I(s)\| \leqslant M/(\sigma - \omega)$, so $I(s)$ is a bounded operator. It is now an elementary calculation to show that

$$I(s)(sI - A)x = x, x \in \mathscr{D}(A)$$

$$(sI - A)I(s)x = x, x \in X$$

Hence $I(s) = (sI - A)^{-1}$. □

The next result characterises the generators of semigroups. Its proof is fairly long and so we shall not give it here.

Theorem 3.2.6 (Hille-Yosida)
In order that a closed linear operator A with dense domain in X generates a strongly continuous semigroup, it is necessary and sufficient that $\exists$ real numbers M,ω such that $\forall$ real $\sigma > \omega$, $\sigma \in \rho(A)$ and

$$\|R(\sigma;A)^m\| \leqslant M(\sigma - \omega)^{-m}, m \geqslant 1 \quad (3.10)$$

Then we have $\|T(t)\| \leqslant Me^{\omega t}$. □
Note that, if $T(t)$ is a contraction semigroup (i.e., $\|T(t)\| \leqslant 1$), then the condition (3.10) takes the form

$$\|R(\sigma;A)^m\| \leqslant \sigma^{-m}, m \geqslant 1$$

Definition 3.2.7
An operator A defined on $\mathscr{D}(A) \subseteq H$ (Hilbert space) is *dissipative* if

$$\text{Re}\langle Ax,x\rangle \leqslant 0, \forall x \in \mathscr{D}(A) \quad (3.11)$$

Theorem 3.2.8 (Phillips and Lumer)
Let A be a linear operator with $\overline{\mathscr{D}(A)} = H$. Then A generates a contraction semigroup iff A is dissipative and $\mathscr{R}(I - A) = H$.

Proof: If Let A be dissipative and let $\sigma > 0$. Then, if $y = \sigma x - Ax (x \in \mathscr{D}(A))$,

$$\sigma \|x\|^2 \leqslant \text{Re}(\sigma\langle x,x\rangle - \langle Ax,x\rangle) = \text{Re}\,\langle y,x\rangle \leqslant \|y\|\cdot\|x\|$$

Since $\mathscr{R}(I - A) = H$, $\sigma = 1 \in \rho(A)$ and $\|R(1;A)\| \leqslant 1$. If $|\sigma - 1| < 1$, then $R(\sigma;A)$ exists, and, by the Neumann series (theorem 2.3.2),

$$R(\sigma;A) = R(1;A)(I + (\sigma - 1)R(1;A))^{-1}$$

$$= R(1;A) \sum_{n=0}^{\infty} ((1 - \sigma)R(1;A))^n$$

and so $\|R(\sigma;A)\| \leqslant \sigma^{-1}$, for $\sigma > 0$ and $|\sigma - 1| < 1$. Again,

$$R(\sigma_1;A) = R(\sigma;A)(I + (\sigma_1 - \sigma)R(\sigma;A))^{-1}$$

for $\sigma_1 > 0$, and $|\sigma_1 - \sigma|\,\|R(\sigma;A)\| < 1$. Hence, as before, $\|R(\sigma_1;A)\| \leqslant \sigma_1^{-1}$. It follows that $\|R(\sigma;A)\| \leqslant \sigma^{-1}$ for all $\sigma > 0$ and so the result follows from the remark after theorem 3.2.6.

Only if Let $T(t)$ be a contraction semigroup. Then,

$$\text{Re}\,\langle T(t)x - x,x\rangle = \text{Re}\,\langle T(t)x,x\rangle - \|x\|^2 \leqslant \|T(t)x\|\cdot\|x\| - \|x\|^2 \leqslant 0$$

Hence, if $x \in \mathscr{D}(A)$

$$\text{Re}\,\langle Ax,x\rangle = \lim_{t\to 0^+} \text{Re}\,\{t^{-1}\langle T(t)x - x,x\rangle\} \leqslant 0. \quad \square$$

Remark 3.2.9
This result can be generalised to Banach spaces using a *semi-inner product*, but we shall not need this result.

Corollary 3.2.10
If $\overline{\mathscr{D}(A)} = H$, then A generates a contraction semigroup if A and A^* are dissipative.

Proof: We must show that $\mathscr{R}(I - A) = H$. Since $(I - A)^{-1}$ is closed and continuous, $\mathscr{R}(I - A)$ is a closed subspace of H. If $\mathscr{R}(I - A) \neq H$, then (by the Hahn-Banach theorem), $\exists 0 \neq x^* \in X^*$ with

$$\langle x - Ax, x^*\rangle = 0 \quad \forall x \in \mathscr{D}(A)$$

Clearly, $x^* \in \mathscr{D}(A^*)$ and then $x^* - Ax^* = 0$, and

$$0 \geqslant \text{Re}\,\langle A^*x^*,x^*\rangle = \langle x^*,x^*\rangle > 0$$

since $x^* \neq 0$. Contradiction. $\square$

We have seen in theorem 3.2.4 that, for any semigroup $T(t)$, $T(t):\mathscr{D}(A) \to \mathscr{D}(A)$. This means that we only obtain a solution of $\dot{x} = Ax$, $x(0) = x_0$ for $x_0 \in \mathscr{D}(A)$. An important class of semigroups has the property that $T(t):X \to \mathscr{D}(A)$, for any T in this class. These are the analytic (or differentiable) semigroups, which we now define via their generators.

Definition 3.2.11
Let A be a closed operator with $\overline{\mathscr{D}(A)} = X$ such that, for some $\phi \in (0,\pi/2)$,

(*a*) $S_{\phi+\pi/2} \triangleq \{\lambda \in \mathscr{C} : |\arg| < \pi/2 + \phi\} \subseteq \rho(A)$, and

(*b*) $\|R(\lambda;A)\| < \dfrac{C}{|\lambda|}$, if $\lambda \in S_\phi$, $\lambda \neq 0$

where C is a constant independent of λ. Then A is called *sectorial*. The following theorem can then be proved.

Theorem 3.2.12
If A is sectorial, then A generates a strongly continuous semigroup $T(t)$ such that

(*a*) $T(t)$ can be continued analytically into the sector

$$S_\phi = \{t \in \mathscr{C} : |\arg t| < \phi, t \neq 0\}$$

(*b*) $AT(t)$ and $\mathrm{d}T(t)/\mathrm{d}t$ are bounded for each $t \in S_\phi$ and

$$\frac{\mathrm{d}T(t)}{\mathrm{d}t} x = AT(t)x, \ x \in X$$

(*c*) For any $0 < \varepsilon < \phi$, $\exists\, C'(= C'(\varepsilon))$ such that

$$\|T(t)\| < C', \ \|AT(t)\| < \frac{C'}{|t|}, \text{ if } t \in S_{\phi-\varepsilon}. \quad \square$$

If A is a positive (i.e. $\langle Ax, x\rangle \geqslant 0$) self-adjoint operator on a Hilbert space, then we can define $A^{1/2}$ in the usual way. In the case of sectorial operators it is also useful to have fractional powers at our disposal.

Definition 3.2.13
If A is sectorial, generates $T(t)$ and $\mathrm{Re}\,\sigma(A) < 0$, then, if $\alpha > 0$, we define

$$A^{-\alpha} \triangleq \frac{1}{\Gamma(\alpha)} \int_0^\infty t^{\alpha-1} T(t)\mathrm{d}t$$

If $\alpha > 0$ we put

$$A^\alpha = (A^{-\alpha})^{-1}$$

It can be shown that A^α has the expected properties, e.g.

(*a*) A^α is closed and densely defined if $\alpha > 0$;
(*b*) if $\alpha \geqslant \beta$, $\mathscr{D}(A^\alpha) \subseteq \mathscr{D}(A^\beta)$;
(*c*) $A^\alpha A^\beta = A^\beta A^\alpha = A^{\alpha+\beta}$ on $\mathscr{D}(A^\gamma)$ with $\gamma = \max(\alpha,\beta,\alpha+\beta)$.

Finally, in this section, we shall state without proof two well-known perturbation results for semigroups.

Theorem 3.2.14
Suppose that $T(t)$ is a strongly continuous semigroup generated by A with $\|T(t)\| < Me^{\omega t}$, and that $B \in \mathscr{B}(X)$. Then $A + B$ is the generator of a strongly continuous semigroup $S(t)$ such that

$$\|S(t)\| < M \exp[(\omega + M\|B\|)t]. \quad \square \tag{3.12}$$

Theorem 3.2.15
Let $T(t)$ be an analytic semigroup with generator A, and let B be a closed operator such that

$$\|Bu\| \leqslant a\|u\| + b\|Au\|, u \in \mathscr{D}(A) \subseteq \mathscr{D}(B) \tag{3.13}$$

(i.e. B is *A-bounded*). Then if a and b are small enough, $A + B$ generates an analytic semigroup. Moreover, if $a = 0$ then we require $b < (1 + C)^{-1}$, where C is as defined in definition 3.2.11(*b*). · □

In particular, if A is sectorial and $a \in \mathscr{R}$, and $A + aI$ generates an analytic semigroup.

3.3 Inhomogeneous evolution equations

We shall now consider the inhomogeneous equation

$$\dot{x}(t) = Ax(t) + f(t), x(0) = x_0 \in X \tag{3.14}$$

where A generates a strongly continuous semigroup $T(t)$ and f is assumed for the moment to be continuously differentiable. If $x(t)$ is known to be a solution of (3.14), then

$$\begin{aligned}\frac{\mathrm{d}}{\mathrm{d}s}(T(t-s)x(s)) &= -AT(t-s)x(s) + T(t-s)\frac{\mathrm{d}x(s)}{\mathrm{d}s}\\ &= T(t-s)f(s)\end{aligned}$$

and since T is strongly continuous we may integrate to obtain

$$x(t) = T(t)x_0 + \int_0^t T(t-s)f(s)\mathrm{d}s \tag{3.15}$$

This relation is called the *variation of constants formula* and shows that the solution of (3.14) is uniquely determined. It is easy to show that $x(t)$ given by (3.15) does, in fact, solve (3.14). One can also weaken the assumption on f as follows.

Theorem 3.3.1
If $T(t-s)f(s) \in \mathscr{D}(A)$ for almost all $t > s \in [0, T]$, $AT(t-s)f(s) \in L^1(]0,t[;X)$ and $f \in L^1(]0,T[;X)$, then $x(t)$ given by (3.15) is the unique solution of (3.14) (a.e.). □

In the case of sectorial operators, it is possible to obtain more precise results. We begin with a definition of a certain class of forcing functions f.

Definition 3.3.2
Let $f:\mathscr{R} \times X \to Y$ for some Banach spaces X,Y. Then f is said to be *locally Holder continuous* in $t \in \mathscr{R}$ and *locally Lipschitz* in $x \in X$ if for each $(t_0,x_0) \in \mathscr{R} \times X$, $\exists$ a neighbourhood U of (t_0,x_0) such that

$$\|f(t,x) - f(s,y)\|_Y \leqslant C_1|t-s|^\theta + C_2\|x-y\|_X$$

for all (t,x), $(s,y) \in U$, and some constants C_1, C_2, $\theta > 0$. In particular, if f is independent of t (or x) then we obtain the obvious separate conditions with $C_2 = 0$ (or $C_1 = 0$).

Definition 3.3.3
If A is sectorial on a Banach space X and Re $\sigma(A) < 0$ we denote by X^α the Banach space $\mathscr{D}(A^\alpha)$ with the graph norm

$$\|x\|_\alpha = \|A^\alpha x\|, \; x \in X^\alpha$$

It is clear that if A has compact resolvent, then the injection $X^\alpha \subseteq X^\beta$ ($\alpha > \beta \geqslant 0$) is compact. The following result can be proved quite easily.

Theorem 3.3.4
If A is sectorial in X, generates $T(t)$, and $f:(0,T) \to X$ is a locally Holder continuous function satisfying

$$\int_0^\rho \|f(t)\| \mathrm{d}t < \infty$$

for some $\rho > 0$, then there exists a unique solution (everywhere in t) of (3.14) given by (3.15). □

We shall now consider the nonlinear equation

$$\frac{\mathrm{d}x}{\mathrm{d}t} = Ax + f(t,x), \; x(t_0) = x_0, \; t > t_0 \tag{3.16}$$

As above, if f is locally Holder continuous in t and locally Lipschitz in x and

$$\int_{t_0}^{t_0+\rho} \|f(s,x(s))\| \mathrm{d}s < \infty, \text{ some } \rho > 0$$

then $x(t)$ satisfies (3.16) iff it satisfies the integral equation

$$x(t) = T(t)x_0 + \int_0^t T(t-s)f(s,x(s))\mathrm{d}s \tag{3.17}$$

We shall now prove that (3.17) has a unique solution, by a method which is applied to many existence problems in the theory of differential equations.

Theorem 3.3.5
Let A be a sectorial operator with Re $\sigma(A) < 0$ which generates a semigroup $T(t)$, and let $0 \leqslant \alpha < 1$. If $f:U \to X$ where U is open in $\mathcal{R} \times X^\alpha$, is locally Holder continuous in t and locally Lipschitz in x, then for any $(t_0,x_0) \in U$, $\exists T > 0$ such that (3.17) has a unique solution on $(t_0,t_0 + T)$ with $x(t_0) = x_0$ (given).

Proof: We shall use the contraction mapping theorem, which states that if $F:M \to M$ is a map in a complete metric space such that

$$\delta(F(m_1), F(m_2)) \leqslant \lambda\delta(m_1,m_2)$$

for each $m_1, m_2 \in M$ and some $\lambda < 1$ (where δ is the metric on M), then F has a fixed point, i.e. $\exists m^* \in M$ such that

$$F(m^*) = m^*$$

The first thing to do in our case is to define M. Let $\beta > 0$, $T > 0$ and define M to be the set of continuous functions $y:[t_0,t_0 + T] \to X^\alpha$ such that $\|y(t) - x_0\|_\alpha \leqslant \beta$ for $t \in [t_0,t_0 + T]$, together with the sup norm

$$\|y\|_M^{\beta,T} = \sup_{t\in[t_0,t_0+T]} \|y(t)\|_\alpha$$

Then M is a complete metric space.
Now choose $\beta > 0$ and $\tau > 0$ such that

$$V \triangleq \{(t,x): t_0 \leqslant t \leqslant t_0 + \tau, \|x - x_0\|_\alpha \leqslant \beta\} \subseteq U$$

and

$$\|f(t,x_1) - f(t,x_2)\| \leqslant C\|x_1 - x_2\|_\alpha$$

some $C > 0$ $((t,x_1), (t,x_2) \in V)$. Put

$$K = \max_{t\in[t_0,t_0+\tau]} \|f(t,x_0)\| \ (< \infty)$$

We must now define the map F for which we want to find a fixed point. This is clearly given by

$$F(y)(t) = T(t - t_0)x_0 + \int_{t_0}^t T(t-s)f(s,y(s))\mathrm{d}s,\ y \in M$$

Now choose $T > 0$ such that $0 < T \leqslant \tau$ and

$$\|(T(h) - I)x_0\|_\alpha \leqslant \beta/2 \text{ for } 0 \leqslant h \leqslant T$$

$$\bar{M}(K + C\beta) \int_0^t u^{-\alpha}\mathrm{d}u \leqslant \beta/2$$

where $\|A^\alpha T(t)\| \leqslant \bar{M}t^{-\alpha}$, $t > 0$ (that the latter inequality holds for some $\bar{M}$ is easy to see).

If $y \in M$, then

$$\begin{aligned}\|F(y)(t) - x_0\|_\alpha &\leqslant \|(T(t - t_0) - I)x_0\|_\alpha + \int_{t_0}^t \|A^\alpha T(t - s)\|\cdot(K + C\beta)\mathrm{d}s \\ &\leqslant \beta/2 + \bar{M}(K + C\beta) \int_{t_0}^t (t - s)^{-\alpha}\mathrm{d}s \\ &\leqslant \beta, \qquad \text{for } t_0 \leqslant t \leqslant t_0 + T\end{aligned}$$

Since $F(y)(t)$ is clearly continuous in t, F maps M into M.

Finally, if $x_1, x_2 \in M$, then

$$\begin{aligned}\|F(x_1)(t) - F(x_2)(t)\|_\alpha &\leqslant \int_{t_0}^t \|A^\alpha T(t - s)\|\cdot\|f(s,x_1(s)) - f(s,x_2(s))\|\mathrm{d}s \\ &\leqslant \bar{M}C \int_{t_0}^t (t - s)^\alpha\mathrm{d}s.\ \|x_1 - x_2\|_M^{\beta,T}\end{aligned}$$

and so

$$\|F(x_1) - F(x_2)\|_M^{\beta,T} \leqslant \tfrac{1}{2}\|x_1 - x_2\|_M^{\beta,T}, \quad x_1,x_2 \in M. \qquad \square$$

One can also prove the usual maximal extension results for these local solutions (see Henry, 1981). We prove finally in this section a result which is useful in stability analysis by using the invariance principle (see Chapter 4).

Theorem 3.3.6

Let the assumptions of the last theorem hold and moreover suppose that A has compact resolvent and f maps $\mathcal{R}_+ \times B \subseteq U \subseteq \mathcal{R} \times X^\alpha$, with B closed and bounded, into bounded sets in X. Then, if $x(t;t_0,x_0)$ is a solution of (3.17) on (t_0,∞) with $\|x(t;t_0,x_0)\|_\alpha$ bounded as $t \to \infty$, $\{x(t;t_0,x_0)\}_{t \geqslant t_0}$ is compact in X^α.

Proof: If $\alpha < \beta < 1$, the inclusion $X^\beta \subseteq X^\alpha$ is compact and so it is enough to show that $\|x(t;t_0,x_0)\|_\beta$ is bounded for $t \geqslant t_0 + 1$. Since A has compact resolvent and Re $\sigma(A) < 0$, $\exists\delta > 0$ such that Re $\sigma(A) < \delta < 0$ and so

$$\begin{aligned}\|x(t;t_0,x_0)\|_\beta &\leqslant \bar{M}(t - t_0)^{-(\beta-\alpha)}\mathrm{e}^{-\delta(t-t_0)}\|x_0\|_\alpha \\ &\quad + \bar{M}C \int_{t_0}^t (t - s)^{-\beta}\mathrm{e}^{-\delta(t-s)}\mathrm{d}s \qquad (3.18)\end{aligned}$$

where we have used the inequality $\|f(t,x(t;t_0,x_0))\| \leqslant C$ for all $t \geqslant t_0$ (some $C > 0$). The right-hand side of (3.18) is clearly bounded for $t \geqslant t_0 + 1$. $\square$

3.4 Nonlinear semigroup theory

For some systems it is not desirable to write them in the semilinear form of (3.16), but to solve the system directly in terms of a nonlinear semigroup. We shall consider, therefore, the most basic results of nonlinear semigroup theory in this section; the proofs may all be found in Barbu (1976).

Consider a system

$$\dot{x} \in Ax, \quad x(0) = x_0 \in C \subseteq H$$

where A is a (multivalued) nonlinear operator defined on a closed subset C of a Hilbert space H. (Note that some of the following results are true in a Banach space, but we shall need only the Hilbert space versions.) We consider a set C which is not necessarily a linear space, since A is not linear and so will not, in general, have a linear space as its domain.

Definition 3.4.1

Let C be closed in H. A *semigroup of contractions* on C is a function S from $[0, \infty[\ xC$ into C such that

(*a*) $S(t + s)x = S(t)S(s)x, \quad \forall x \in C,\ t,s \geqslant 0$
(*b*) $S(0)x = x, \quad \forall x \in C$
(*c*) $S(t)x$ is continuous in $t \geqslant 0, \quad \forall x \in C$
(*d*) $\|S(t)x - S(t)y\| \leqslant \|x - y\|, \quad \forall t > 0,\ \forall\, x,y \in C$

The *generator* A of $S(t)$ is defined in formally the same way as (3.7) (with S replacing T and C replacing $\mathcal{D}(A)$).

If S satisfies (*a*), (*b*), (*c*) and

$$\|S(t)x - S(t)y\| \leqslant e^{\omega t}\|x - y\|$$

then S is called a *semigroup of type* ω.

Definition 3.4.2

If H is a real Hilbert space we say that a nonlinear (multivalued) operator A is *dissipative* if

$$\langle y_1 - y_2, x_1 - x_2 \rangle_H \leqslant 0, \quad \forall x_1,x_2 \in \mathcal{D}(A),\ y_i \in Ax_i$$

A is *maximal (or m-) dissipative* if A has no proper dissipative extension in H.

Lemma 3.4.3

Let C be a closed convex subset of a real Hilbert space H and let A be a dissipative operator in H with $0 \in \mathcal{D}(A) \subseteq C$. Then there exists a maximal dissipative operator $\bar{A}$ with $\mathcal{D}(A) \subseteq \mathcal{D}(\bar{A}) \subseteq C$ which extends A. □

Theorem 3.4.4
Let S be a semigroup of nonlinear contractions defined on a closed subset C of H and let A be the generator of S. Then, if $x \in \mathscr{D}(A)$, we have

(a) $S(t)x \in \mathscr{D}(A)$ $\forall t \geqslant 0$ and the function $t \rightarrow AS(t)x$ is right continuous.
(b) $S(t)x$ has right derivative $(\mathrm{d}/\mathrm{d}t)^+S(t)x$ $\forall t \geqslant 0$, and $(\mathrm{d}/\mathrm{d}t)^+S(t)x = AS(t)x$, $\forall t \geqslant 0$.
(c) $\dfrac{\mathrm{d}}{\mathrm{d}t}S(t)x = AS(t)x$ for almost all $t \geqslant 0$. □

In view of (c) in theorem 3.4.4, we say that a function $x(t)$ satisfies the equation

$$\dot{x} = Ax$$

on $[0, \infty[$ if $x(t)$ is continuous in t, Lipschitz on compact intervals of $[0, \infty[$ and $x(t)$ is differentiable a.e. on $]0, \infty[$ with

$$\frac{\mathrm{d}x}{\mathrm{d}t}(t) = Ax(t) \text{ a.e. } t > 0$$

(Contrast this with the linear case.)

Theorem 3.4.5
Let A be a maximal dissipative set in HxH. Then there exists a unique semigroup of contractions $S(t)$ on $\overline{\mathscr{D}(A)}$ with generator A^0, where $A^0x = \min \|y\|$, with $y \in Ax$. □
(This is a partial generalisation of the Hille-Yosida theorem to nonlinear semigroups. There is also a converse which we shall not discuss).

Note finally that A generates a nonlinear semigroup of type ω if $A - \omega I$ is m-dissipative.

3.5 Examples

In this section we shall present a number of examples of systems to which we shall refer later in the book. These systems will be written here in the form of their natural dynamics (without control); the control term will be considered later.

Example 3.5.1
The simplest parabolic system is the heat conduction equation (with constant thermal conductivity normalised to 1 for simplicity) given by

$$\frac{\partial T}{\partial t} = \frac{\partial^2 T}{\partial x^2}, \; T(0,t) = T(1,t) = 0, \; T(x,0) = T_0(x) \tag{3.19}$$

(This is the equation of heat diffusion in a one-dimensional bar of unit length.

The three-dimensional equation in general domains can be considered similarly, but an explicit expression for the semigroup is harder to obtain.) From the results of Chapter 2 and Section 3.2 it follows that we may write (3.19) in the form

$$\frac{d\phi}{dt} = A\phi,\ \phi(t)(x) = T(x,t),\ \phi(0) = T_0(x) \tag{3.20}$$

where A is a sectorial operator defined in $L^2([0,1])$ by

$$(A\phi)(x) = \frac{d^2\phi}{dx^2},\ \phi(0) = \phi(1) = 0 \tag{3.21}$$

with domain $H^2([0,1]) \cap H_0^1([0,1])$.

By the separation of variables it is easy to see that the semigroup $T(t)\phi(0)$ generated by A is given by

$$T(t)\phi(0) = \sqrt{2}\sum_{n=1}^{\infty} \phi_n(0)e^{-n^2\pi^2 t}\sin(n\pi x) \tag{3.22}$$

where

$$\phi_n(0) = \sqrt{2}\,\langle\phi(0), \sin n\pi x\rangle_{L^2([0,1])}$$

Similarly, if we take the boundary conditions $\phi_x(0) = \phi_x(1) = 0$ in (3.21) we obtain the semigroup

$$T(t)\phi(0) = \langle\phi(0),1\rangle_{L^2} + \sum_{n=1}^{\infty} 2e^{-n^2\pi^2 t}\cos n\pi x\,\langle\phi(0), \cos n\pi x\rangle_{L^2} \tag{3.23}$$

(The domain of A is now $\mathscr{D}(A) = \{\phi \in H^2([0,1]): \phi_x(0) = \phi_x(1) = 0\}$.)

Example 3.5.2

The following nonlinear diffusion equation is of interest in nuclear reactor dynamics (Kastenberg, 1969):

$$\frac{\partial\phi}{\partial t} = \frac{\partial^2\phi}{\partial x^2} + \lambda\phi - \rho\phi^2,\ \mu,\lambda \geqslant 0,\ \lambda < \pi^2,\ x \in [0,1] \tag{3.24}$$

Consider the operator A defined in $L^2([0,1])$ by

$$A\phi = \frac{\partial^2\phi}{\partial x^2} + \lambda\phi - \rho\phi^2,\ \phi \in \mathscr{D}(A) = H^2([0,1]) \cap H_0^1([0,1])$$

and let

$$C = \{\phi \in H_0^1([0,1]): \phi(x) \geqslant 0,\ \forall x \in [0,1]\}$$

Now, if $\phi_1, \phi_2 \in C \cap \mathscr{D}(A)$, then, with $\phi = \phi_1 - \phi_2$

$$\begin{aligned}\langle A\phi_1 - A\phi_2, \phi_1 - \phi_2\rangle_{L^2([0,1])} &= -\|\phi'\|^2_{L^2} + \lambda\|\phi\|^2_{L^2} - \rho \int_0^1 (\phi_1 + \phi_2)\phi^2 dx \\ &\leqslant -\|\phi'\|_{L^2} + \lambda\|\phi\|_{L^2} \\ &\leqslant (\lambda - \pi^2)\|\phi\|_{L^2}\end{aligned}$$

by the well-known inequality

$$\pi^2\|\phi\|^2_{L^2} \leqslant \|\phi'\|^2_{L^2}, \ \forall\phi \in H^1_0([0,1])$$

It follows from theorem 3.3.5 that the system has a solution on C (by the maximum principle; Friedman, 1964), and that the solution exists for all time $t \geqslant 0$. Also A generates a nonlinear semigroup of type $\lambda - \pi^2$ on C.

Example 3.5.3

A hyperbolic equation which has received much attention in the control literature is the wave equation, the simplest one-dimensional version being

$$\frac{\partial^2 h(x,t)}{\partial t^2} = \frac{\partial^2 h(x,t)}{\partial x^2} - \alpha \frac{\partial h(x,t)}{\partial t}, \quad x \in [0,1] \tag{3.25}$$

with $h(0,t) = h(1,t) = 0$, $h(x, 0) = h_0(x)$, $\alpha \geqslant 0$. The term involving α corresponds to some form of damping.

Let A be the operator defined by (3.21) and put

$$\phi(t)(x) = h(x,t)$$

$$\psi(t)(x) = \frac{\partial h(x,t)}{\partial t}$$

Then we obtain the system

$$\frac{d}{dt}\begin{pmatrix}\phi \\ \psi\end{pmatrix} = \begin{pmatrix}0 & I \\ A & -\alpha\end{pmatrix}\begin{pmatrix}\phi \\ \psi\end{pmatrix} \triangleq \mathscr{A}\begin{pmatrix}\phi \\ \psi\end{pmatrix} \tag{3.26}$$

We shall use corollary 3.2.10 to show that $\mathscr{A}$ generates a semigroup. First note that $\mathscr{D}((-A)^{1/2}) = H^1_0([0,1])$, A is dissipative on $L^2([0,1])$ and $(-A)^{1/2}$ is positive and self-adjoint.

Now introduce the Hilbert space $H = H^1_0([0,1]) \oplus L^2([0,1])$ with inner product

$$\langle(\phi_1,\psi_1), (\phi_2,\psi_2)\rangle_H = \langle(-A)^{1/2}\phi_1, (-A)^{1/2}\phi_2\rangle_{L^2} + \langle\psi_1,\psi_2\rangle_{L^2}$$

Then, we have

$$\begin{aligned}\langle(\phi,\psi), \mathscr{A}(\phi,\psi)\rangle_H &= \langle(-A)^{1/2}\phi, (-A)^{1/2}\psi\rangle_{L^2} + \langle\psi, -A\phi - \alpha\psi\rangle_{L^2} \\ &= \langle A\phi,\psi\rangle + \langle\psi,-A\phi - \alpha\psi\rangle \\ &= -\alpha\|\psi\|^2\end{aligned}$$

where it is clear that $\mathscr{D}(\mathscr{A}) = (H_0^1([0,1]) \cap H^2([0,1]) \oplus H_0^1([0,1])$, (i.e. $\mathscr{D}(A) \oplus \mathscr{D}((-A)^{1/2})$). Similarly, it can be shown that

$$\mathscr{A}^* = \begin{pmatrix} 0 & -I \\ A & -\alpha \end{pmatrix}$$

satisfies $\langle(\phi,\psi), \mathscr{A}^*(\phi,\psi)\rangle_H = -\alpha\|\psi\|^2$ and so the conditions of corollary 3.2.10 are satisfied.

If $\alpha = 0$, it follows by separation of variables that $\mathscr{A}$ generates the group $T(t)$ given by

$$T(t)\begin{pmatrix} \phi \\ \psi \end{pmatrix} = \begin{pmatrix} \Sigma 2\{\langle\phi,e_n\rangle_{L^2}\cos n\pi t + \dfrac{1}{n\pi}\langle\psi,e_n\rangle_{L^2}\sin n\pi t\}e_n \\ \Sigma 2\{-n\pi\langle\phi,e_n\rangle_{L^2}\sin n\pi t + \langle\psi,e_n\rangle_{L^2}\cos n\pi t\}e_n \end{pmatrix} \tag{3.27}$$

for $\begin{pmatrix} \phi \\ \psi \end{pmatrix} \in H$, with $e_n(x) = \sin n\pi x$. Note that the inner product on H is simply

$$\langle(\phi_1,\psi_1), (\phi_2,\psi_2)\rangle_H = \int_0^1 (d/dx)\phi_1(d/dx)\phi_2 dx + \int_0^1 \psi_1\psi_2 dx$$

Example 3.5.4

The semigroup theory above can also be applied to delay equations of the form

$$\dot{x}(t) = A_0x(t) + \sum_{i=1}^{N} A_i\begin{cases} x(t+\theta_i); & t+\theta_i \geq 0 \\ h(t+\theta_i); & t+\theta_i < 0 \end{cases}$$

$$+ \int_{-b}^{0} A_{01}(\theta)\begin{cases} x(t+\theta)d\theta; & t+\theta \geq 0 \\ h(t+\theta)d\theta; & t+\theta < 0 \end{cases}$$

$x(0) = h(0)$, $-b \leqslant \theta_n < \theta_{n-1} < \ldots < \theta_1 < 0, A_0, A_i \in \mathscr{B}(H)$, $A_{01} \in C(-b,0;\mathscr{B}(H))$

In fact, it can be shown (Delfour and Mitter, 1972) that we obtain the semigroup

$$(T(t)h)(\theta) = \begin{cases} x(t+\theta); & t+\theta \geq 0 \\ h(t+\theta); & t+\theta < 0 \end{cases}$$

on

$$\mathscr{M}^2(-b,0;H) \triangleq H \times L^2(-b,0;H)$$

However, rather than discuss the general case further, we shall consider the simplest delay equation (*cf.* Banks and Abbasi-Ghelmansarari, 1983a)

$$\begin{aligned} \dot{x}(t) &= x(t-\delta) \\ x(t) &= f(t),\ t \in [-\delta,0),\ x(0) = x_0 \end{aligned} \tag{3.28}$$

where f is (real) analytic on $(-\infty,\infty)$. If $x(t)$ is any solution of (3.28) on $[-\delta,\infty)$, we extend x to $(-\infty,\infty)$ by defining

$$x(t) = f^{(i)}(t),\ t \in [-(i+1)\delta,\ -i\delta),\ i \geqslant 0$$

Now introduce the new state $\boldsymbol{x} = (x_1,x_2,\ldots)^T \in l^2$ by

$$x_i(t) = x(t-(i-1)\delta),\ t \in (-\infty,\infty),\ i \geqslant 1 \tag{3.29}$$

Then it is easy to see that

$$\dot{\boldsymbol{x}}(t) = A\boldsymbol{x}(t),\ t \in [0,\delta) \tag{3.30}$$

is equivalent on the interval $[0,\delta)$ to the delay equation (3.28), with initial condition

$$\boldsymbol{x}(0) = (x_0, f(-\delta), f'(-\delta), f''(-\delta),\ldots)^T \tag{3.31}$$

where A is the left-shift operator on l^2 defined in example 2.3.5. Although we only have equivalence of (3.28) and (3.30) on $[0,\delta)$ we can also show that the solution of (3.28) can be obtained from

$$\boldsymbol{x}(t) = \exp\{A(t-i\delta)\}[P_i\boldsymbol{x}(i\delta) + (A^*)^i\boldsymbol{x}(0)]$$

for $t \in [i\delta,(i+1)\delta)$ by using the definition (3.29), where P_i is the projection onto $\{\boldsymbol{x} \in l^2\colon \boldsymbol{x} = (x_1,\ldots,x_i,0,0,\ldots)\}$ and A^* is the dual of A (i.e. the right-shift operator).

We therefore see that the system (3.30), although not equivalent to (3.28), has useful properties which we shall exploit in the study of stability of (3.28). Also, the system (3.30) will easily show some interesting properties of the infinite-dimensional root locus.

Example 3.5.5

In plasma physics, the first-order approximation to the Boltzmann equation is called the Vlasov equation and is used to model the ion and electron densities in a collisionless plasma. If these densities are denoted respectively by f_{1i} and f_{1e} then Vlasov's equation is

$$\frac{\partial f_{1j}}{\partial t} + \frac{\boldsymbol{v}\partial f_{1j}}{\partial \boldsymbol{r}} + \frac{q_j}{m_j c}\,\boldsymbol{v} \times \boldsymbol{B}\,\frac{\partial f_{1j}}{\partial \boldsymbol{v}} + \frac{q_j}{m_j}\,\boldsymbol{E}_\mathrm{p}\,\frac{\partial f_{0j}}{\partial \boldsymbol{v}} = 0, \quad j = i \text{ or } e \tag{3.32}$$

where $\boldsymbol{r} = (x_1,x_2,x_3)$ is the spatial coordinate, $\boldsymbol{v}$ is the velocity, $\boldsymbol{B}$ is the magnetic field, $\boldsymbol{E}_\mathrm{p}$ is the plasma electric field, f_{0j} is the appropriate equilibrium distribution and q_j, m_j, c are constants. If we consider the equations

$$\frac{dx_1}{v_1} = \frac{dx_2}{v_2} = \frac{dx_3}{v_3} = \frac{dv_1}{\alpha_j(v_2B_3 - v_3B_2)} = \frac{dv_2}{\alpha_j(v_3B_1 - v_1B_3)}$$
$$= \frac{dv_3}{\alpha_j(v_1B_2 - v_2B_1)} = dt$$

with $\alpha_j = q_j/m_jc$, $\boldsymbol{B} = (B_1,B_2,B_3)^T$ and write them in the form

$$\dot{z} = \boldsymbol{A}_j z$$

where $z = (x_1,x_2,x_3,v_1,v_2,v_3)^T \in \mathcal{R}^6$ and

$$\boldsymbol{A}_j = \begin{bmatrix} \boldsymbol{0}_3 & \boldsymbol{I}_3 \\ \boldsymbol{0}_3 & \alpha_j \begin{bmatrix} 0 & B_3 & -B_2 \\ -B_3 & 0 & B_1 \\ B_2 & -B_1 & 0 \end{bmatrix} \end{bmatrix}$$

then we easily see that the operator

$$\mathsf{A}_j = -\left(\boldsymbol{v}\frac{\partial}{\partial \boldsymbol{r}} + \alpha_j \boldsymbol{v} \times \boldsymbol{B}\frac{\partial}{\partial \boldsymbol{v}}\right)$$

generates a semigroup $T_j(t)$ on $L^2(\mathcal{R}^6)$ defined by

$$(T_j(t)f)(z) = f(\exp(-\boldsymbol{A}_j t)z),\ t \geq 0,\ f \in L^2(\mathcal{R}^6)$$

It now follows from theorem 3.2.14 that, if we recall the relation

$$\boldsymbol{E}_p = 4\pi \sum q_j \int_\Omega \nabla_r \frac{1}{\|\boldsymbol{r} - \boldsymbol{r}'\|} \int_{V_c} f_{1j} d\boldsymbol{v} d\boldsymbol{r}'$$

(where Ω is the region containing the plasma and $V_c = \{\boldsymbol{v}: \|\boldsymbol{v}\| \leq c\}$, c = speed of light) then the complete operator

$$A \triangleq -\begin{pmatrix} \boldsymbol{v}\frac{\partial}{\partial \boldsymbol{r}} + \alpha_i \boldsymbol{v} \times \boldsymbol{B}\frac{\partial}{\partial \boldsymbol{v}} + \frac{q_i}{m_i}\boldsymbol{E}_p\frac{\partial f_{0i}}{\partial \boldsymbol{v}} \\ \boldsymbol{v}\frac{\partial}{\partial \boldsymbol{r}} + \alpha_e \boldsymbol{v} \times \boldsymbol{B}\frac{\partial}{\partial \boldsymbol{v}} + \frac{q_e}{m_e}\boldsymbol{E}_p\frac{\partial f_{0e}}{\partial \boldsymbol{v}} \end{pmatrix}$$

generates a semigroup $T(t)$ on $L^2(\mathcal{R}^6) \oplus L^2(\mathcal{R}^6)$. (For more details on this example see Boyd and Sanderson, 1969; Banks and Mousavi-Khalkhali, 1982).

Example 3.5.6
As a final example, we shall consider the linear multipass process (Edwards and Owens, 1981; Banks, 1981a; 1982)

$$\frac{dx_k(t)}{dt} = A_0x_k(t) + A_1x_{k-1}(t) + \ldots + A_lx_{k-l}(t) \tag{3.33}$$

for $k \geqslant 0$, $t \in [0,\tau]$, with memory of length l. Such equations describe many real engineering processes and have received some attention recently in the literature. In the present case, the state $x_k(t)$ belongs to a Hilbert space H, A_0 is an (unbounded) operator in H with domain $\mathscr{D}(A_0)$, and A_i, $1 \leqslant i \leqslant l$ are bounded operators on H (the case of unbounded operators A_i can be treated similarly). The data $x_{-1}(t),\ldots,x_{-l}(t)$, $t \in [0,\tau]$ and $x_k(0)$, $k \geqslant 0$ is specified, and τ is called the pass length of the process.

We introduce the Hilbert spaces

$$\mathscr{H}_m = \bigoplus_{i=0}^{m} H \qquad (m \geqslant l)$$

and define the operator $\mathscr{A}_m$ by the matrix representation (of order $(m+1) \times (m+1)$)

$$\mathscr{A}_m \triangleq \begin{bmatrix} A_0 & & & & & & & \\ A_1 & A_0 & & \underline{0} & & & & \\ \cdots\cdots\cdots & & & & & & & \\ A_l & A_{l-1} & \cdots & A_0 & & & & \\ 0 & A_l & A_{l-1} & \cdots & A_0 & & & \\ \cdots\cdots\cdots\cdots\cdots\cdots\cdots\cdots & & & & & & & \\ 0 & \cdots & 0 & A_l & A_{l-1} & \cdots & A_0 & \end{bmatrix}$$

Then, if A_0 generates a semigroup $T(t)$, it can be shown (Banks, 1982) that $\mathscr{A}_m$ also generates a semigroup $\mathscr{T}_m(t)$ on $\mathscr{H}_m$ given by

$$\mathscr{T}_m(t) = (S_{ij}(t))_{0 \leqslant i \leqslant m, 0 \leqslant j \leqslant m} \tag{3.34}$$

with

$$S_{ij}(t) = 0 \qquad \text{if } j > i$$

$$S_{ii}(t) = T(t), \text{ if } 0 \leqslant i \leqslant m$$

and

$$S_{ij}(t) = \sum_{k=2}^{j-i} \sum_{I_k} K_k^{i_1 \ldots i_{k-1}}(t,0), \text{ if } i > j$$

where

$$I_k = \{(i_1,\ldots,i_{k-1}): j - i - 1 = i_1 + \ldots + i_{k-1}, i_n \in \{1,\ldots,l\}, 0 \leqslant n \leqslant k-1\}$$

and the kernels K are defined inductively by

$$K_1(t,s) = T(t-s)$$

$$K_{k+1}^{i_1\ldots i_k}(t,s) = \int_s^t K_k^{i_1\ldots i_{k-1}}(t,s_1)A_{i_k}T(s_1 - s)\mathrm{d}s_1 \qquad (3.35)$$

Chapter 4

Stability theory

4.1 Introduction

One of the most important aspects of dynamical systems theory is that of stability. The techniques of stability analysis are well known in the finite dimensional situation, and the basic ideas generalise in most cases (with some important qualifications) to infinite dimensional systems, as we shall see. The classical methods can be conveniently divided into state-space (or time-domain) and frequency-domain techniques. However, when dealing with nonlinear systems which contain linear subsystems, a mixture of both methods is employed – the input–output theory.

We shall now list the finite dimensional methods which we shall generalise in this chapter to distributed systems; it is assumed that the reader is familiar with these results (see, for example: Lyapunov, 1949; LaSalle and Lefschetz, 1961; Hahn, 1956).

(*a*) *Linear systems:* $\dot{x} = Ax$, $A \in \mathscr{B}(\mathscr{R}^n,\mathscr{R}^n)$

(i) The system is globally asymptotically stable iff $\sigma(A) \subseteq \mathscr{C}^0_-$.

(ii) The system is globally asymptotically stable iff $\exists$ a positive definite Hermitian matrix P for each such matrix Q such that

$$A^*P + PA = -Q$$

(Lyapunov's theorem).

(*b*) *Nonlinear systems*

(i) *Linearisation*: If the system $\dot{x} = f(x)$ has an isolated equilibrium point at $x = 0$ and we write

$$\dot{x} = Ax + G(x)x$$

where A such that $\sigma(A) \subseteq \mathscr{C}^0_-$, then 0 is an asymptotically stable equilibrium point of the system.

(ii) *Circle theorem* (Sandberg, 1964; Zames, 1966): Consider the nonlinear system shown in Fig. 4.1, where $u(t) = \psi\,[f(t),t]$ and ψ has the properties

(1) $\psi(0,t) = 0,\ t \geqslant 0$

(2) $\exists \alpha,\beta \in \mathcal{R},\ \beta > 0$ such that

$$\alpha \leqslant \frac{\psi(x,t)}{x} \leqslant \beta, \quad t \geqslant 0$$

Also

$$K(s) = \int_0^\infty k(t)\mathrm{e}^{-st}\mathrm{d}t,\ s = \sigma + \mathrm{i}\omega,\ -\infty < \omega < \infty$$

for some function k. Then the system is L^2-stable (i.e. $g \in L^2$ implies $v \in L^2$) if one of the following conditions holds:

(1) $\alpha > 0$ and the locus of $K(\mathrm{i}\omega)$ for $-\infty < \omega < \infty$ lies outside the circle C of radius $(\alpha^{-1} - \beta^{-1})/2$ with centre $(-\frac{1}{2}(\alpha^{-1} + \beta^{-1}),0)$ and does not encircle C.
(2) $\alpha = 0$ and $\mathrm{Re}[K(\mathrm{i}\omega)] > -\beta^{-1}$ for all real ω.
(3) $\alpha < 0$ and the locus of $K(\mathrm{i}\omega)$ is contained in the circle of radius $\frac{1}{2}(\alpha^{-1} - \beta^{-1})$ with centre $(-\frac{1}{2}(\alpha^{-1} + \beta^{-1}),0)$.

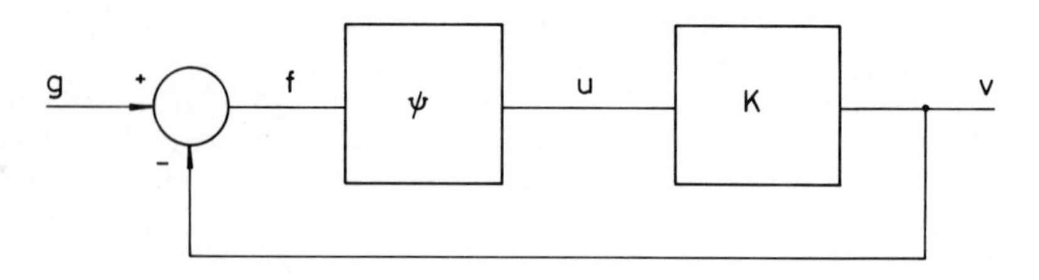

Fig. 4.1 *Nonlinear feedback system*

If the system is not scalar, but has $f, u, v, g \in L^2([0,\infty);\mathcal{R}^n)$ then the same result holds with the conditions (1)–(3) replaced by

(1′) $\det\,[I_n + \frac{1}{2}(\alpha + \beta)K(s)] \neq 0$, for $\sigma \geqslant 0$

(2′) $\frac{1}{2}(\alpha - \beta) \sup_{-\infty<\omega<\infty} \Lambda\{[I_n + \frac{1}{2}(\alpha + \beta)K(\mathrm{i}\omega)]^{-1}K(\mathrm{i}\omega)\} < 1$

where

$$\Lambda\{M\} \triangleq \sup_{\lambda\in\sigma(M^*M)} \sqrt{\lambda}. \quad \square$$

(iii) *Lyapunov's main stability theorem*: Consider the system defined by the differential equation

$$\dot{x} = f(x,t)$$

and suppose that the origin $x = 0$ is an isolated equilibrium point of the system, i.e. $f(0,t) = 0 \forall t \geqslant 0$. If there exists a (*Lyapunov*) function $V(x,t)$ with continuous first partial derivatives such that

(1) $V(x,t)$ is positive definite

(2) $\dfrac{dV}{dt} = \text{grad } V.f + \dfrac{\partial V}{\partial t}$ is negative definite

(3) $V(x,t) \to \infty$ as $\|x\| \to \infty$

uniformly in t, then the origin is uniformly asymptotically stable in the large. □

Of course, there are many more classical results on stability but we shall only discuss the distributed versions of the above theorems. To begin, we shall give precise definitions of stability in the next section and introduce the notion of dynamical system.

4.2 Definitions of stability

We shall consider basically two kinds of stability in this book, Lyapunov stability and input–output stability, and discuss their relevance in systems theory. We start by defining a dynamical system in order to introduce LaSalle's invariance principle later in the chapter.

Definition 4.2.1
Let C be a closed subset of a complete metric space. A *dynamical system* on C is a family of maps $S = \{S(t):t \geqslant 0\}$ such that the following conditions hold:

(*a*) For each t, $S(t):C \to C$ is a continuous map.
(*b*) For each $x \in C$, $t \to S(t)x$ is continuous.
(*c*) $S(0) = I_C$.
(*d*) $S(t)S(\tau) = S(t + \tau)$, $\forall t, \tau \geqslant 0$.

Of course, a dynamical system on a Banach space is just a (nonlinear) semigroup. If S is a dynamical system on C, the *positive semi-orbit through x* is just the set

$$\gamma(x) = \{S(t)x:t \geqslant 0\}$$

(i.e. the 'solution' of the system with initial condition x). If $\gamma(x)$ consists of just the single point x we say that x is an *equilibrium point*, and if $\gamma(x)$ is such that there is a minimum positive $p > 0$ for which $S(p)x = x$ then we say that γ is periodic with period p.

Definition 4.2.2
Let S be a dynamical system on $C \subseteq (M, \rho)$ (a metric space with metric ρ).

Then we say that an orbit $\gamma(x)$ is *stable* if for each $\varepsilon > 0$ there exists $\delta(= \delta(\varepsilon)) > 0$ such that

$$\rho(x,y) < \delta \ (y \in C) \rightarrow \rho(S(t)x, S(t)y) < \varepsilon$$

The orbit γ is *unstable* if it is not stable. γ is *uniformly asymptotically stable* if it is stable and $\exists r > 0$ such that

$$\rho(x,y) < r \ (y \in C) \rightarrow \rho(S(t)y, S(t)x) \rightarrow 0 \text{ as } t \rightarrow \infty$$

uniformly in y.

Definition 4.2.3
A *Lyapunov function* for the dynamical system S is a continuous, real-valued function V on C such that

(*a*) $V(0) = 0$
(*b*) $V(x) \geqslant \alpha(\|x\|) \quad (\|x\| = \rho(x,0))$
where $\alpha \in CI_0$ (the space of continuous strictly increasing real-valued functions f with $f(0) = 0$)

$$(c)\ \dot{V}(x) = \limsup_{t \rightarrow 0+} \frac{1}{t} \{V(S(t)x) - V(x)\} \leqslant 0$$

Many real engineering systems do not appear naturally as differential equations, but must be regarded as a 'black box' into which one can input information and from which one can obtain output data. It is therefore very convenient to have a definition of stability which is directly applicable to these cases. Consider, then, a system Σ (as in Fig. 4.2) which maps the vector space $M([0,\infty);X)$ (of measurable maps from $[0,\infty)$ into a Banach space X) into itself. (One can, of course, work on the smaller space $L^p_e([0,\infty);X)$ – the *extended* L^p space.) The system Σ will be defined by some (nonlinear) map $N{:}M([0,\infty);X) \rightarrow M([0,\infty);X)$. Now, N maps the Banach space $L^p([0,\infty);X)$ into $M([0,\infty);X)$. However, there is no guarantee, *a priori*, that N will map L^p into any other such space. Hence, we introduce the following definition.

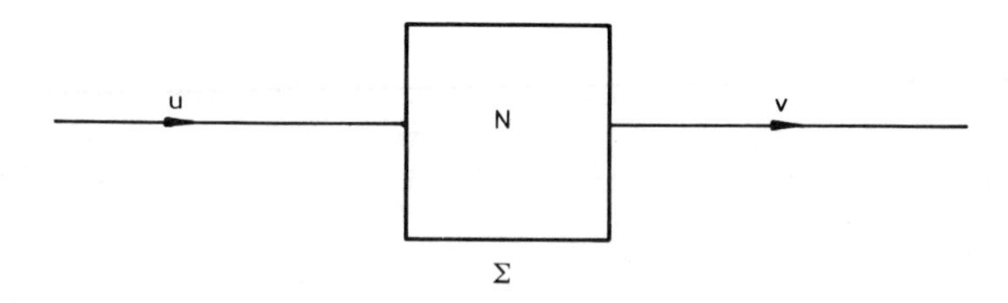

Fig. 4.2 *Input–output system* Σ

Definition 4.2.4
The system Σ is *(p,q) input–output stable* (or just (p,q)-stable) if

$$N(L^p([0,\infty);x)) \subseteq L^q([0,\infty);X) \tag{4.1}$$

Slightly more generally, it may be that at each time t, the output $v(t)$ belongs not to X, but to another Banach space Y; then (4.1) becomes

$$N(L^p([0,\infty);X)) \subseteq L^q([0,\infty);Y)$$

The context should make clear to which spaces the inputs or outputs belong (see Desoer and Vidyasagar, 1975; Banks and Collingwood, 1979).

4.3 Stability of homogeneous linear systems

In this section we consider the linear system

$$\dot{x} = Ax, \quad x(0) = x_0 \tag{4.2}$$

where A generates a semigroup $T(t)$ on a Banach space X. We note immediately that the result (a)(i) of Section 4.1 is not true in general. Indeed, the classical example of Hille and Phillips (1957) shows that there is a semigroup $U(t)$ with generator which has empty spectrum although $\|U(t)\| = e^{\pi t}, t \geqslant 0$. It is therefore clear that in order to obtain a result of this type, one must impose further conditions on A. Also, it is convenient to introduce the following definition of stability which is stronger than asymptotic stability.

Definition 4.3.1
The system (4.2) is *exponentially stable* if there exists positive constants M,ω such that

$$\|T(t)\| \leqslant Me^{-\omega t}, \ t \geqslant 0$$

Clearly, exponential stability implies asymptotic stability, but the converse is not generally true. In order to generalise (a)(i) of Section 4.1 we define

$$\omega = \sup \{\text{Re } \lambda : \lambda \in \sigma(A)\}$$

and

$$\bar{\omega} = \inf \{\omega : \|T(t)\| \leqslant Me^{\omega t}, \text{ for some } M \geqslant 0 \text{ and all } t \geqslant 0\}$$

It follows that $\omega < \bar{\omega}$, by proposition 3.2.5. The equality $\omega = \bar{\omega}$ is known in the following cases:

(a) A is bounded;
(b) $T(t)$ is an analytic semigroup;
(c) $T(t)$ is compact for some $t_1 > 0$ (and hence all $t \geqslant t_1$).

Clearly case (b) implies case (a). We shall prove the equality $\omega = \bar{\omega}$ for analytic semigroups, the proof of (c) may be found in the literature (Zabczyk, 1976).

Proof of (b): This follows from theorem 3.2.12. First note that we can suppose that

$$\sup \operatorname{Re} \sigma(A) = 0 \tag{4.3}$$

for, if not, we just replace A by $A - aI$ for some $a \in \mathcal{R}$. Then, by theorem 3.2.12, if (4.3) holds, we have

$$\|T(t)\| \leqslant C$$

for some constant C, and so the assumption $0 = \omega < \bar{\omega}$ leads directly to a contradiction. Hence, $\omega = \bar{\omega}$. □

The equality $\omega = \bar{\omega}$ is known as the *spectrum determined growth condition*. Hence, we have seen that, in the cases (a), (b) and (c) above, a linear system is exponentially stable iff it has its spectrum contained in a closed subset of $\mathscr{C}^0_-$, i.e.

$$\sup \operatorname{Re} \sigma(A) < 0$$

It is also easy to see that if $\sup \operatorname{Re} \sigma(A) > 0$ then the system is (Lyapunov) unstable. However, the behaviour of the system when $\sigma(A) \cap \{\lambda : \operatorname{Re} \lambda = 0\} = \phi$ appears to be an open question in general. Of course, if the spectrum on the imaginary axis consists entirely of values in the point spectrum, each of finite multiplicity, then the answer is easy when the eigenvalues are discrete.

We consider now the conditions under which exponential stability is guaranteed by the existence of a solution of Lyapunov's equation as in (b)(i) in Section 4.1. A complete characterisation for Hilbert spaces is given by Datko (1970) and we shall follow his proof, and a simplification due to Pritchard and Zabczyk (1977). First we need a definition.

Definition 4.3.2

A *Hermitian operator* on a complex Hilbert space H is a bounded operator B such that

$$\langle Bx,y\rangle = \overline{\langle By,x\rangle} = \langle x,By\rangle$$

It follows from Riesz and Sz-Nagy (1955) that if $B(t)$ is a monotonically increasing family of Hermitian operators which is bounded above (i.e. $\langle B(t)x,x\rangle \leqslant M\,\|x\|^2$), then there exists a Hermitian operator B such that $B(t)$ converges strongly to B. We first prove the following.

Lemma 4.3.3

Let $T(t)$ be a strongly continuous semigroup with generator A. Then the integral $\int_0^\infty \|T(s)x\|^2 \mathrm{d}s < \infty$, $\forall\, x \in H$ iff there exists a Hermitian operator B on H such that $B \geqslant 0$ and

$$2\text{Re}\,\langle BAx,x\rangle = \langle BAx,x\rangle + \overline{\langle BAx,x\rangle} = \langle BAx,x\rangle + \langle A^*Bx,x\rangle = -\,\|x\|^2 \tag{4.4}$$

Proof: Assume

$$\int_0^\infty \|T(s)x\|^2 \mathrm{d}s < \infty,\ \forall x \in H$$

and define

$$B(t)x = \int_0^t T^*(s)T(s)x\mathrm{d}s,\ t \geqslant 0$$

Clearly, $B(t)$ is Hermitian and

$$|\langle B(t)x,y\rangle|^2 \leqslant \left(\int_0^\infty \|T(s)x\|^2\mathrm{d}s\right)\left(\int_0^\infty \|T(s)y\|^2\mathrm{d}s\right)$$

Hence, by the uniform boundedness principle,

$$\sup_{0\leqslant t<\infty} \|B(t)\| < \infty$$

and since $B(t)$ is nondecreasing (i.e., $0 \leqslant \langle B(t)x,x\rangle \leqslant \langle B(t')x,x\rangle,\ t \leqslant t'$) it follows from the above remark that there exists a Hermitian operator B such that $B(t)$ converges strongly to B. If $x \in \mathscr{D}(A)$, then the map

$$t \rightarrow \langle BT(t)x,T(t)x\rangle = \int_0^\infty \|T(t + s)x\|^2\mathrm{d}s$$

is differentiable and has derivative

$$2\text{Re}\,\langle BAT(t)x,T(t)x\rangle = -\,\|T(t)x\|^2$$

and so the necessity of the condition is proved.

To prove sufficiency suppose that B satisfies the conditions. Then the function

$$V(x,t) = \langle BT(t)x,T(t)x\rangle \geqslant 0$$

satisfies

$$\frac{\mathrm{d}V}{\mathrm{d}t}(x,t) = 2\text{Re}\,\langle BAT(t)x,T(t)x\rangle = -\,\|T(t)x\|^2$$

for $x \in \mathscr{D}(A)$. Hence, integrating,

$$0 \leqslant V(x,t) = V(x,0) - \int_0^t \|T(s)x\|^2\mathrm{d}s$$

and so

$$V(x,0) \geqslant \int_0^t \|T(s)x\|^2\mathrm{d}s,\ \forall t \geqslant 0,\ x \in \mathscr{D}(A) \tag{4.5}$$

Now $T(t)x_n \rightarrow T(t)x$ as $x_n \rightarrow x$ uniformly on compact intervals and so (4.5) is, in fact, true for all $x \in H$. Hence,

$$V(x,0) \geqslant \int_0^\infty \|T(s)x\|^2 \mathrm{d}s, \ \forall x \in H. \quad \square$$

Theorem 4.3.4
The convergence of the integral $\int_0^\infty \|T(t)x\|^2 \mathrm{d}t$ for each $x \in H$ is equivalent to the existence of $M,\omega > 0$ such that

$$\|T(t)\| \leqslant M\mathrm{e}^{-\omega t}$$

Proof: Clearly, the existence of such an M and ω implies that $\int_0^\infty \|T(t)x\|^2 \mathrm{d}t < \infty$. If, conversely, the integral converges then define the bounded operators P_n by

$$\langle P_n x, x \rangle = \int_0^n \|T(t)x\|^2 \mathrm{d}t$$

The sequence $\{P_n\}$ is monotonic and bounded above and by the uniform boundedness theorem has a bounded limit P. Hence

$$\int_0^\infty \|T(t)x\|^2 \mathrm{d}t = \lim_n \langle P_n x, x \rangle = \langle Px, x \rangle \leqslant \|P\| \cdot \|x\|^2$$

Let M_1, ω_1 be such that

$$\|T(t)\| \leqslant M_1 \mathrm{e}^{\omega_1 t}, \ \forall t \geqslant 0$$

Then it is easy to see that

$$\|T(t)x\|^2 \leqslant M_1^2 \|P\| \frac{2\omega}{1 - \mathrm{e}^{-2\omega t}} \|x\|^2, \quad t > 0$$

Hence there exists $\gamma > 0$ such that

$$\|T(t)\| \leqslant \gamma, \ t \geqslant 0$$

and so

$$\|T(t)x\|^2 = \frac{1}{t} \int_0^t \|T(t)x\|^2 \mathrm{d}s \leqslant \frac{1}{t} \int_0^t \|T(s)\|^2 \, \|T(t-s)x\|^2 \mathrm{d}s$$

$$\leqslant \frac{\gamma^2}{t} \|P\| \cdot \|x\|^2$$

Thus, $\|T(t)\| \to 0$ as $t \to \infty$. Hence,

$$\bar{\omega} = \lim \frac{\log \|T(t)\|}{t} < 0$$

(*cf.* theorem 3.2.3) and so there exist $M,\omega > 0$ with

$$\|T(t)\| \leqslant M\mathrm{e}^{-\omega t}. \quad \square$$

Corollary 4.3.5
Exponential stability of a linear system is equivalent to the existence of a Hermitian operator B such that (4.4) holds. □

Example 4.3.6
In the case of the heat equation given in example 3.5.1, it is easy to see that, if T is given by (3.22), then

$$Bx = \int_0^\infty T^*(s)T(s)x\mathrm{d}s = \int_0^\infty T(2s)x\mathrm{d}s$$

$$= \sqrt{2} \sum_{n=1}^{\infty} (x_n/(2n^2\pi^2)) \sin n\pi x$$

where $x_n = \sqrt{2}\,\langle x, \sin n\pi x\rangle$, and

$$\int_0^\infty \|T(s)x\|^2\mathrm{d}s = \sum_{n=1}^{\infty} \left(\frac{x_n}{2n^2\pi^2}\right)$$

4.4 Local stability by linearisation

The method which we describe in this section is discussed in detail in Kirchgassner and Kielhofer (1973) and Henry (1981). We consider now the semilinear system

$$\frac{\mathrm{d}x}{\mathrm{d}t} = Ax + f(x,t) \tag{4.6}$$

where A is such that $A_1 = A + aI$ is sectorial for some $a \in \mathcal{R}$ and $\operatorname{Re} \sigma(A + aI) < 0$. The system is defined on a Banach space X and we introduce the new norm on X by defining $\|x\|_\alpha = \|A_1^\alpha x\|$, $0 < \alpha < 1$ (*cf.* definition 3.3.3). We shall consider the stability of (4.6) in the norm $\|.\|_\alpha$ for an equilibrium point $x_0 \in \mathcal{D}(A)$, i.e. a point where $Ax_0 = -f(x_0,t)$ for all $t \geqslant t_0$.

Theorem 4.4.1
Let A, x_0 be as above and suppose that f is defined in $X^\alpha \times \mathcal{R}$. We assume that f is locally Holder continuous in t, locally Lipschitz in x and can be written in the form

$$f(x_0 + z,t) = f(x_0,t) + Bz + g(z,t)$$

where B is a bounded linear map from X^α into X and $\|g(z,t\| = o(\|z\|_\alpha)$ as $\|z\|_\alpha \to 0$, uniformly in $t \geqslant 0$.

Then, if $\operatorname{Re} \sigma(A + B) < \beta < 0$, x_0 is a uniformly asymptotically stable equilibrium point of (4.6).

Proof: $A + B$ generates an analytic semigroup and so is exponentially stable (since $\operatorname{Re} \sigma(A + B) < 0$), by the results of Section 4.3. Also, if $S(t)$ is the semigroup generated by $A + B$, and $\operatorname{Re} \sigma(A + B) < \beta_1 < \beta$ then

$$\left.\begin{array}{l} \|S(t)z\|_\alpha \leqslant M e^{\beta_1 t}\|z\|_\alpha \\ \|S(t)z\|_\alpha \leqslant M t^{-\alpha} e^{\beta_1 t}\|z\| \end{array}\right\}, \ t > 0, \ z \in X^\alpha$$

for some $M \geqslant 1$. Now let $\sigma > 0$ be chosen small enough so that

$$M\sigma \int_0^\infty s^{-\alpha} e^{(\beta_1 - \beta)s} ds < 1/2$$

and $\rho > 0$ sufficiently small in order that

$$\|g(z,t)\| \leqslant \sigma\|z\|_\alpha \text{ for } \|z\|_\alpha \leqslant \rho, \ t \geqslant 0$$

Put $z(t) = x(t;0,x_1) - x_0$. Then if $\|x_1 - x_0\|_\alpha \leqslant \rho/2M$, the solution of (4.6) will exist and will satisfy

$$\|z(t)\|_\alpha \leqslant \rho$$

for sufficiently small $t > 0$. Suppose that $\|z(t)\|_\alpha < \rho$ on $[0,t_1)$. Then

$$\|z(t)\|_\alpha = \|S(t)z(0) + \int_0^t S(t - s)g(z(s),s)ds\|_\alpha$$

$$\leqslant \rho/2 + \rho\sigma M \int_{-\infty}^t (t - s)^{-\alpha} e^{\beta_1(t-s)} ds < \rho, \ t \in [0,t_1)$$

Hence, we must, in fact, have $\|z(t)\|_\alpha < \rho$, $t \in [0,\infty)$. Also, it follows easily from the above inequalities that

$$\|z(t)\|_\alpha \leqslant 2Me^{\beta_1 t}\|x_1 - x_0\|_\alpha \quad \square$$

4.5 Input–output stability and the circle theorem

Input–output stability of a single (open-loop) system has been defined in Section 4.2. Consider now the feedback system shown in Fig. 4.3, where the systems S_1 and S_2 are, in general, nonlinear operators defined in the following spaces

$$\left.\begin{array}{l} S_1 : L^{p_1}(0,t;X_1) \to L^{p_2}(0,t;X_2) \\ S_2 : L^{p_2}(0,t;X_2) \to L^{p_1}(0,t;X_1) \end{array}\right\} \text{ for each } t \geqslant 0 \qquad \text{for each } t \geqslant 0 \quad (4.7)$$

for some Banach spaces X_1 and X_2. Clearly, we must also require

$$u_i, \ e_i \in L^{p_i}(0,t;X_i), \ i = 1,2$$

$$y_1 \in L^{p_2}(0,t;X_2), \ y_2 \in L^{p_1}(0,t;X_1), \text{ for each } t \geqslant 0$$

The basic result of input–output stability theory is the following 'small-gain theorem' (Sandberg, 1964; Zames, 1966; Banks and Collingwood, 1979).

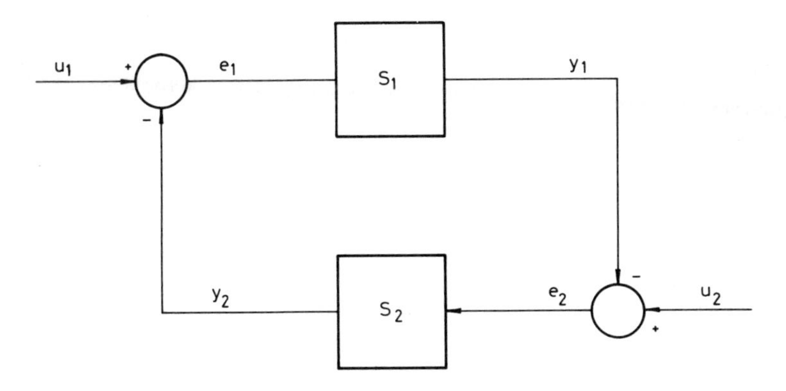

Fig. 4.3 *General feedback system with subsystems S_1,S_2*

Theorem 4.5.1
Define the gains γ_i of S_i $(i = 1,2)$ by

$$\gamma_1 = \sup_{t>0} \sup_{x \neq 0} \|S_1 x\|_{L^{p_2}(0,t;X_2)} / \|x\|_{L^{p_1}(0,t;X_1)} \tag{4.8}$$

and similarly for γ_2. If each γ_i is finite and we have

$$\gamma_1 \gamma_2 < 1$$

then the system in Fig. 4.3 is input–output stable in the sense that

$$\begin{pmatrix} u_1 \\ u_2 \end{pmatrix} \in L^{p_1}(0,\infty;X_1) \oplus L^{p_2}(0,\infty;X_2) \rightarrow$$

$$\begin{pmatrix} y_1 \\ y_2 \end{pmatrix} \in L^{p_2}(0,\infty;X_2) \oplus L^{p_1}(0,\infty;X_1)$$

Proof: The proof is simple. We just note that

$$y_1 - S_1 c_1 - S_1(u_1 - y_2) = S_1(u_1 - S_2(u_2 - y_1))$$

and so, for any $t \geqslant 0$,

$$\begin{aligned} \|y_1\|_{L^{p_2}(0,t;X_2)} &\leqslant \gamma_1 \|u_1 - S_2(u_2 - y_1)\|_{L^{p_1}(0,t;X_1)} \\ &\leqslant \gamma_1 \|u_1\|_{L^{p_1}(0,t;X_1)} + \gamma_1 \|S_2(u_2 - y_1)\|_{L^{p_1}(0,t;X_1)} \\ &\leqslant \gamma_1 \|u_1\|_{L^{p_1}(0,t;X_1)} + \gamma_1 \gamma_2 \|u_2 - y_1\|_{L^{p_2}(0,t;X_2)} \end{aligned}$$

and so, if $\gamma_1 \gamma_2 < 1$, then

$$\|y_1\|_{L^{p_2}(0,t;X_2)} \leqslant (1 - \gamma_1 \gamma_2)^{-1} \{\gamma_1 \|u_1\|_{L^{p_1}(0,t;X_1)} + \gamma_1 \gamma_2 \|u_2\|_{L^{p_2}(0,t;X_2)}\}$$

If we let $t \to \infty$, then we have $y_1 \in L^{p_2}(0,\infty;X_2)$; a similar proof shows that $y_2 \in L^{p_1}(0,\infty;X_1)$. □

Remark 4.5.2
The small gain theorem also holds in the case when the nonlinear systems S_i are multivalued; i.e. when, for example S_1 is a relation, which means that $S_1 \subseteq L^{p_1}(0,t;X_1) \times L^{p_2}(0,t;X_2)$ for each $t \geqslant 0$. The definition (4.8) of the gain of S_1 is easily generalised to this case.

Remark 4.5.3
The space

$$L_e^{p_1}(0,\infty;X_1) = \bigcup_{t \geqslant 0} L^{p_1}(0,t;X_1)$$

with the inductive limit topology is usually called the extended space of $L^{p_1}(0,\infty;X_1)$. The conditions (4.7) on S_1 and S_2 merely state that each should map between the appropriate extended spaces.

Example 4.5.4
Consider the system defined by the nonlinear diffusion equation

$$\frac{\partial \phi}{\partial t} = \frac{\partial^2 \phi}{\partial x^2} + f(\phi(x,t),t) + u(x,t) \tag{4.9}$$

where $x \in [0,1]$, $\phi(x,0) = \phi_0(x)$, $\phi(0,t) = \phi(1,t) = 0$ and u is an input function. Let the operator A be defined by

$$A\phi = \frac{d^2\phi}{dx^2}, \ \phi \in \mathscr{D}(A) = H^2([0,1]) \cap H_0^1([0,1])$$

and let $T(t)$ be the semigroup generated by A. Then we can write equation (4.9) in the form

$$\bar{\phi}(t) = T(t)\phi_0 + \int_0^t T(t-s)(f(\bar{\phi}(s),s) + \bar{u}(s))\mathrm{d}s \tag{4.10}$$

where $\bar{\phi}(t)(x) = \phi(x,t)$, $\bar{u}(s)(x) = u(x,s)$. This equation is in the form of the feedback system in Fig. 4.3 if we define S_1 and S_2 by

$$S_1(\phi)(x,t) = f(\phi(x,t),t)$$

$$S_2(\psi)(x,t) = \int_0^t [T(t-s)\psi(s)](x)\mathrm{d}s$$
$$= (T^*\psi)(t)$$

and

$$u_1 = T(t)\phi_0, \ u_2 = \bar{u}, \ e_1 = \bar{\phi}$$

Suppose that $S_1:L^r(0,t;L^2[0,1]) \to L^q(0,t;L^2[0,1])$ and has finite gain $\gamma_1 = G(S_1) < \infty$. Now recall, from (3.22), that

$$\|T(t)\|_{\mathscr{L}(L^2[0,1])} \leqslant e^{-\pi^2 t} \tag{4.11}$$

Now, by corollary 2.7.2, we have

$$\|S_2\psi\|_{\mathscr{L}^r([0,t];L^2[0,1])} \leqslant \|T\|_{L^p([0,\infty);(L^2[0,1]))}\|\psi\|_{L^q([0,t];L^2[0,1])}$$

for any $p,q,r \geqslant 1$ such that $r^{-1} = p^{-1} + q^{-1} - 1$, and for all $t > 0$. However,

$$\|T\|_{L^p([0,\infty);\mathscr{L}(L^2[0,1]))} \leqslant \left(\int_0^\infty (e^{-\pi^2 t})^p \mathrm{d}t\right)^{1/p} = (p\pi^2)^{-1/p}$$

and $u_1 = T(t)\phi_0 \in L^r([0,\infty);L^2[0,1])$ for any $\phi_0 \in L^2[0,1]$. Hence, if $\bar{u} \in L^q([0,\infty);L^2[0,1])$, then theorem 4.5.1 shows that the system is input–output stable in the (q,r)-sense if

$$\gamma_1 < (p\pi^2)^{1/p}$$

Remark 4.5.5

If the input u in the above example is operating pointwise, i.e. $u(x,t) = \delta(x - x_1)g(t)$ for some $x_1 \in [0,1]$ and some function g, then one can obtain similar results to those above by noting that

$$\|T(t)\phi\|_{H^{1/2+\varepsilon}[0,1]} \leqslant \frac{Me^{-\pi^2 t}}{t^{1/4+1/2\varepsilon}}\|\phi\|_{L^2[0,1]}$$

and dually,

$$\|T(t)\|_{L^2[0,1]} \leqslant \frac{Me^{-\pi^2 t}}{t^{1/4+1/2\varepsilon}}\|\phi\|_{H^{-1/2-\varepsilon}[0,1]}$$

Now we recall that $\delta \in H^{-1/2-\varepsilon}[0,1]$ (*cf.* Section 2.5).

We shall now discuss the generalisation of the circle theorem to certain distributed systems. The main result will not be proved here, a simple proof being given by Banks (1981*b*), based on the finite dimensional result of Sandberg (1964). We first introduce some notation. Let B be a bounded normal operator on a Hilbert space H (i.e. $B^*B = BB^*$). Then it can be shown (Yosida, 1974) that

$$r(B) \triangleq \sup\{|\lambda|:\lambda \in \sigma(B)\} = \|B\|$$

For any bounded linear operator M on H, we write

$$\Lambda(M) \triangleq \{r(M^*M)\}^{1/2}$$

Finally, if $S \subseteq \mathscr{C}$, we define the ε-neighbourhood of S, denoted $N_\varepsilon(S)$ by

$$N_\varepsilon(S) = \{c \in \mathscr{C}: \exists s \in S \text{ with } |c - s| < \varepsilon\}$$

We now state the circle theorem.

Theorem 4.5.6
Let $T(t)$ be a stable semigroup and let

$$g(t) = x(t) + \int_0^t T(t - s)f(x(s))\mathrm{d}s$$

where $g \in L^2(0,\infty;H)$ and f satisfies the following condition.
There exist numbers a,b with $a < b$ such that

$$\|f \circ h - \alpha h\|_{L^2(0,\infty;H)} \leqslant \eta(\alpha)\|h\|_{L^2(0,\infty;H)}$$

for any real α with

$$\eta(\alpha) = \max\,[(\alpha - a), (b - \alpha)]$$

Let A be the generator of $T(t)$ and suppose that:

(*a*) $0 \in N_\varepsilon(\sigma(I + \tfrac{1}{2}(a + b)R(s;A)))$ for $\operatorname{Re} s \geqslant 0$, and some $\varepsilon > 0$;
(*b*) $\tfrac{1}{2}(b - a) \sup\limits_{\omega} \Lambda\{[I + \tfrac{1}{2}(a + b)R(\mathrm{i}\omega;A)]^{-1}R(\mathrm{i}\omega;A)\} < 1$.

Then

$$x \in L^2(0,\infty;H). \quad \square$$

(Recall that $R(s;A) = (sI - A)^{-1}$ for $s \notin \sigma(A)$.) In order to interpret this result in terms of a nonencirclement criterion note first that

$$\lambda \in \sigma(A) \text{ iff } \frac{1}{s - \lambda} \in \sigma(R(s;A))$$

Now let $\xi_s: \mathscr{C}\backslash\{s\} \rightarrow \mathscr{C}$ be the function defined by

$$\xi_s(\lambda) = 1/(s - \lambda)$$

and assume that A is self-adjoint. By the spectral mapping theorem,

$$\sigma(I + \tfrac{1}{2}(a + b)R(s;A)) = 1 + \tfrac{1}{2}(a + b)\xi_s(\sigma(A)) \tag{4.12}$$

and since $R(\mathrm{i}\omega;A)^* = R(-\mathrm{i}\omega;A)$, we have

$$\tfrac{1}{2}(b - a)\sigma\{R(-\mathrm{i}\omega;A)[I + \tfrac{1}{2}(a + b)R(-\mathrm{i}\omega;A)]^{-1}[I + \tfrac{1}{2}(a + b)R(\mathrm{i}\omega;A)]^{-1}R(\mathrm{i}\omega;A)\}$$

$$= \{\tfrac{1}{2}(b - a)|(1 + \tfrac{1}{2}(a + b)\xi_{\mathrm{i}\omega}(\lambda))^{-1}\xi_{\mathrm{i}\omega}(\lambda)|^2 : \lambda \in \sigma(A)\}$$

assuming $0 \notin \sigma(I + \tfrac{1}{2}(a + b)R(\mathrm{i}\omega;A))$, $-\infty < \omega < \infty$.
By (4.12), condition (*a*) is satisfied if, for each $\lambda \in \sigma(A)$,

$$0 \notin N_\varepsilon(1 + \tfrac{1}{2}(a + b)\xi_s(\lambda)) \text{ for } \operatorname{Re} s \geqslant 0$$

However, as is well known, this is true if the polar plot of $\xi_{i\omega}(\lambda)$ does not encircle or enter the disc $N_\varepsilon([-2(a+b)^{-1},0])$, and so (a) is satisfied if

$$\left.\begin{array}{l}\text{The region traced out by the set valued map } \omega \to \xi_{i\omega}(\sigma(A)) \text{ does} \\ \text{not contain a curve which encircles or enters the disc } N_\varepsilon([-2(a \\ + b)^{-1},0]), \text{ for some } \varepsilon > 0.\end{array}\right\} \quad (4.13)$$

Similarly, condition (b) above is satisfied if

$$\left.\begin{array}{l}\text{The region traced out by the map } \omega \to \xi_{i\omega}(\sigma(A)) \text{ does not} \\ \text{intersect the region } R(a) \subseteq \mathscr{C} \text{ for } -\infty < \omega < \infty, \text{ where} \\ \text{(i) } R(a) = \text{disc of radius } \tfrac{1}{2}(a^{-1} - b^{-1}) \text{ with centre } [-\tfrac{1}{2}(a^{-1} + \\ b^{-1}),0] \text{ if } a > 0. \\ \text{(ii) } R(a) = \text{half-plane Re } s \leqslant -b^{-1} \text{ if } a = 0. \\ \text{(iii) } R(a) = \text{exterior of the disc in (i) if } a < 0.\end{array}\right\} \quad (4.14)$$

Example 4.5.7
Consider the system

$$\dot{z}(t) = Az(t) - f(z(t)) \tag{4.15}$$

where A is self-adjoint and generates a stable semigroup which is analytic, and f satisfies

$$\langle f(h(x)) - ah(x), f(h(x)) - bh(x)\rangle_{L^2(0,1)} \leqslant 0 \tag{4.16}$$

Then, $\sigma(A)$ lies in a sector

$$S_{d,\phi} = \{\lambda : \pi - \phi < |\arg(\lambda - d)| < \pi, \lambda \neq d\}, \; d < 0$$

It is easy to show that $\xi_{i\omega}(\lambda)$ maps $S_{d,\phi}$ into the circle of radius $1/2d$ and centre $(1/2d,0)$ for any $\omega \in (-\infty,\infty)$. Hence, if $a \geqslant 0$, or $a < 0$ and $2d > -a$, then the solution z of (4.15) belongs to $L^2(0,\infty;H)$. In particular, consider the system

$$\frac{\partial z}{\partial t} = \frac{\partial^2 z}{\partial x^2} + pz - f(z(t,x)), \; x \in [0,1] \tag{4.17}$$

with boundary conditions $\partial z/\partial x = 0$ when $x = 0$ or 1; the operator

$$Az = \frac{d^2 z}{dx^2} + pz$$

with domain

$$\mathscr{D}(A) = \{z \in L^2(0,1): \frac{d^2 z}{dx^2} \in L^2(0,1), \frac{dz}{dx} = 0 \text{ at } x = 0,1\}$$

generates an analytic semigroup and has the spectrum

$$\sigma(A) = \{\lambda : \lambda = -p - (j-1)^2\pi^2, j \geqslant 1\}$$

Hence, if (4.16) holds and $a \geqslant 0$ or $a < 0$ and $2p > -a$, then the solution z of (4.17) belongs to $L^2(0,\infty;L^2(0,1))$. Note that (4.16) implies that

$$\|f \circ z - \tfrac{1}{2}(a + b)z\| < \tfrac{1}{2}(b - a)\|z\|$$

A more general version of theorem 4.5.6 is given by Freedman *et al*. (1969), and a generalisation of the Nyquist criterion to certain distributed systems is proved by Desoer and Wang (1980).

4.6 Lyapunov stability

In the final section of this chapter we shall consider Lyapunov stability of distributed systems. The definition of stability and Lyapunov functions has been given in Section 4.2. The first result is proved in a similar manner to the analogous finite dimensional result (see for example, Yoshizawa, 1966).

Theorem 4.6.1
Let $\{T(t), t \geqslant 0\}$ be a dynamical system on a set C contained in a Banach space X, and suppose that V is a Lyapunov function for this dynamical system, defined on C, such that $\dot{V}(x) \leqslant -\beta(\|x\|)$, $\beta \in CI_0$. Then the origin 0 in X is an asymptotically stable equilibrium point of $T(t)$. □

Example 4.6.2
Consider the system

$$\frac{\partial \phi}{\partial t} = \frac{\partial^2 \phi}{\partial x^2} + \lambda\, \phi - \rho\, \phi^2 \tag{4.18}$$

discussed in example 3.5.2, and define

$$V = \|x\|^2_{L^2(0,1)}$$

(It has been shown that this system defines a dynamical system on $C = \{\phi \in H_0^1[0,1]: \phi \geqslant 0\}$.) Now,

$$\dot{V} = \overline{\langle x,x\rangle}^{\,\cdot}_{L^2} = 2\,\langle \dot{x},x\rangle_{L^2} = 2\,\langle Ax,x\rangle_{L^2}$$

$$\leqslant (\lambda - \pi^2)\|x\|_{L^2}$$

as shown previously. Hence, if $\lambda < \pi^2$, $\phi = 0$ is an asymptotically stable equilibrium point of (4.18), in the L^2 sense.

Example 4.6.3
Consider the system

$$\frac{\partial \phi}{\partial t} = \frac{\partial^2 \phi}{\partial x^2} - \phi^3, x \in [0,1], \phi(0,t) = \phi(1,t) = 0 \tag{4.19}$$

Define the operator A by

$$A\phi = \frac{d^2\phi}{dx^2} - \phi^3, \quad \phi \in H^2[0,1] \cap H_0^1[0,1]$$

Then we have

$$\langle A\phi_1 - A\phi_2, \phi_1 - \phi_2 \rangle_{L^2} = -\|\phi_1' - \phi_2'\|_{L^2} - \int_0^1 (\phi_1^3 - \phi_2^3)(\phi_1 - \phi_2)dx$$

$$\leqslant -\pi^2 \|\phi_1 - \phi_2\|_{L^2} - \int_0^1 (\phi_1 - \phi_2)^2(\phi_1^2 + \phi_1\phi_2 + \phi_2^2)dx$$

$$\leqslant -\pi^2 \|\phi_1 - \phi_2\|_{L^2}$$

for each $\phi \in \mathscr{D}(A)$ and so (4.19) defines a dynamical system $S(t)\phi$ on $H_0^1[0,1]$, and $S(t)\phi \to 0$ in $L^2[0,1]$ as $t \to \infty$ for any $\phi \in H_0^1[0,1]$. (Consider again the function $V = \|\phi\|_{L^2}^2$.) We can, however, even show that $S(t)\phi \to 0$ in $H_0^1[0,1]$ as follows. Consider the function

$$V(\phi) = \int_0^1 \{\tfrac{1}{2}\phi'^2 + \tfrac{1}{4}\phi^4\}\, dx \geqslant \tfrac{1}{2}\|\phi\|_{H_0^1}^2$$

Then,

$$\dot{V}(\phi) = \int_0^1 \{\phi_x \phi_{xt} + \phi^3 \phi_t\} dx$$

$$= \int_0^1 \{-\phi_{xx}\phi_t + \phi^3 \phi_{xx} - \phi^6\} dx$$

$$= \int_0^1 \{-\phi_{xx}^2 + 2\phi^3 \phi_{xx} - \phi^6\} dx$$

However,

$$\int_0^1 \phi^3 \phi_{xx} dx = -\int_0^1 3\phi_x^2 \phi^2 dx$$

and so

$$\dot{V} = \int_0^1 \{-\phi_{xx}^2 - 6\phi_x^2\phi^2 - \phi^6\} dx \leqslant -\pi^2 \int_0^1 \phi_x^2 dx$$

whence $S(t)\phi \to 0$ in $H_0^1[0,1]$.

In the last part of this chapter we shall discuss briefly the invariance principle of LaSalle (see Hale, 1969). Let $S(t)$ be a dynamical system on a subset C of a Banach space X.

Definition 4.6.4
A subset $K \subseteq C$ is *invariant* (for $S(t)$) if, for any $x_0 \in K$, there exists a continuous curve $x:\mathcal{R} \to K$ such that $x(0) = x_0$ and

$$S(t)x(\tau) = x(t+\tau) \text{ for all } \tau \in (-\infty,\infty),\ t \geqslant 0$$

Definition 4.6.5
If $x_0 \in C$ and $\gamma(x_0) = \{S(t)x_0 : t \geqslant 0\}$ is the orbit through x_0, then the ω*-limit set* of x_0 is

$$\omega(x_0) \stackrel{\Delta}{=} \{x \in C : \exists t_n \to \infty \text{ such that } S(t_n)x_0 \to x\}$$

The two basic theorems below are proved in Henry (1981).

Theorem 4.6.6
Let $x_0 \in C$ and $\gamma(x_0) \subseteq$ a compact set in C. Then $\omega(x_0)$ is nonempty, compact, invariant and connected; moreover

$$\inf_{y \in \omega(x_0)} \|S(t)x_0 - y\| \to 0 \text{ as } t \to \infty. \quad \square$$

Theorem 4.6.7
If V is a Lyapunov function on C and we define

$$E \stackrel{\Delta}{=} \{x \in C : \dot{V}(x) = 0\}$$

and M to be the maximal invariant set in E, then if $\gamma(x_0)$ lies in a compact subset of C, we have $\omega(x_0) \subseteq E$, and

$$S(t)x_0 \to M \text{ as } t \to \infty. \quad \square$$

Example 4.6.8
Consider again the system (4.18) with $\lambda = \pi^2$, $\rho > 0$. We know that this equation defines a local dynamical system on $H_0^1[0,1]$. Let A be the operator $A = -\mathrm{d}^2/\mathrm{d}x^2$ with $\mathscr{D}(A^{1/2}) = H_0^1[0,1]$, and consider the function

$$V(\phi) = \int_0^1 \{(\phi'(x))^2 - \pi^2\phi^2(x) + \tfrac{1}{2}\, a\phi^4(x)\}\mathrm{d}x,\ \phi \in H_0^1[0,1]$$

Then

$$\dot{V}(\phi) = -2\int_0^1 \{\phi''(x) + \pi^2\phi(x) - a\phi^3(x)\}^2\mathrm{d}x$$

for $\phi \in \mathscr{D}(A)$. Hence V is a Lyapunov function, and

$$V(\phi) \geqslant \frac{a}{2}\int_0^1 \phi^4 \mathrm{d}x$$

Therefore, along any solution $\phi(x,t)$ we have

$$V(\phi(x,0)) \geqslant V(\phi(x,t)) \geqslant \frac{a}{2}\int_0 \phi^4(x,t)\mathrm{d}x$$

and so $\int_0^1 \phi^4 \mathrm{d}x$ is bounded. Thus,

$$\int_0^1 \phi^2 \mathrm{d}x < (\int_0^1 \phi^4 \mathrm{d}x)^{1/2}$$

is bounded and by the definition of V

$$\|\phi(.,t)\|_{H_0^1[0,1]}$$

is bounded. Hence, we have a dynamical system on $H_0^1[0,1]$ with all orbits bounded. However, A has compact resolvent, so each orbit is precompact, and by theorem 4.6.7,

$$\phi(.,t) \rightarrow \omega(\phi(.,0)) \subseteq \{\phi : \dot{V}(\phi) = 0\}$$

in $H_0^1[0,1]$, as $t \rightarrow \infty$. But $\dot{V}(\phi) = 0$ implies that

$$\phi'' + \pi^2\phi - a\phi^3 = 0 \text{ on } [0,1], \ \phi(0) = \phi(1) = 0$$

Hence,

$$a\int_0^1 \phi^4 \mathrm{d}x = \int_0^1 (\phi'' + \pi^2\phi)\phi \mathrm{d}x = \int_0^1 \{-(\phi')^2 + \pi^2\phi^2\}\mathrm{d}x \leqslant 0$$

i.e. $\phi = 0$

4.7 Notes

In this chapter we have given a brief outline of the theory of stability of distributed parameter systems. There are, of course, many more results on the stability of such systems than are presented here. For example, the comparison principle may be used to study certain problems in combusion and population genetics as discussed by Aronson (1978). Further results on the generation of Lyapunov functions are given by Pao (1968), Buis (1968), and Walker (1976). Finally, we should also mention infinite dimensional versions of the Hopf bifurcation which can be found in Holmes and Marsden (1981) and Crandall and Rabinowitz (1977).

Chapter 5

Controllability, observability and realisation theory

5.1 Introduction

In the next chapter we shall discuss the theory of optimal control of distributed systems, in which one determines the 'best' control from a class of controls which achieves some performance objective. Before doing this it is clearly desirable to know under which conditions at least one control exists which will drive the system from one state to another. This is the theory of controllability, which we shall now discuss, along with the related concepts of observability and stabilisability and we shall also mention the spectral assignability problem.

Consider first the familiar results in finite dimensional systems theory. Let

$$\begin{aligned} \dot{x} &= Ax + bu \\ y &= Cx \end{aligned} \tag{5.1}$$

be a finite dimensional system with state $x \in \mathcal{R}^n$ and control $u \in \mathcal{R}^m$, with matrices A, B and C of appropriate sizes. The system (or the pair (A, B)) is *(completely) controllable* if for any $x_1, x_2 \in \mathcal{R}^n$ there exists a control function u which transfers x_1 to x_2 in some finite time. It is well known that (5.1) is controllable if and only if the matrix

$$[B \quad AB \quad A^2B \quad \dots \quad A^{n-1}B]$$

is of full rank. The system (or (A, B)) is *(completely) observable* if from a knowledge of the control $u(t)$ and the output $y(t)$ over some interval $[0,T]$, one can determine the initial state $x(t_0)$. It is well known that (5.1) is observable if and only if the matrix

$$[C^T \quad A^TC^T \quad (A^T)^2C^T \quad \dots \quad (A^T)^{n-1}C^T]$$

is of full rank.

Frequently, one requires only to stabilise the system (5.1) with a feedback control law of the form $u = Dx$. The system (5.1) is called *stabilisable* if there exists a matrix D such that $\sigma(A + BD)$ is contained in the open left half-plane.

Note that stabilisability $\leftarrow$ controllability (Wonham, 1979) and that stabilisability $\rightarrow$ controllability of the unstable modes. Also, it can be shown (Russell, 1974) that if (A,B) is stabilisable for both positive and negative times (i.e. $A + BD_1$, $-(A + BD_2)$ are stable matrices for some (D_1,D_2) then (A,B) is controllable, and the same proof holds for distributed systems.

We may also require a certain degree of stability in the system together with appropriate damping and we then study the pole allocation problem; i.e., does there exist a real matrix D such that $\sigma(A + BD)$ is any prescribed set of at most n complex values? It is known (Wonham, 1979) that for $m = 1$ the solubility of this problem is equivalent to the controllability of the pair (A,B). This is easily seen by reducing the system to the 'control canonical form'

$$\dot{x} = \begin{bmatrix} 0 & 1 & & & \\ & 0 & 1 & 0 & \\ & & & & \\ & & & & \\ & 0 & & 0 & 1 \\ -a_n & -a_{n-1} & \dots & & -a_1 \end{bmatrix} x + \begin{bmatrix} 0 \\ 0 \\ \cdot \\ \cdot \\ \cdot \\ 0 \\ 1 \end{bmatrix} u \tag{5.2}$$

Finally, in this chapter, we discuss the important theory of linear infinite dimensional realisations of input–output maps. Of course, a map

$$T(t):[0,\infty) \rightarrow \mathcal{R}$$

is realisable by a finite dimensional linear system (A,b,c) (i.e. (5.1)) if $T(t)$ is of exponential order (i.e. $|T(t)| \leqslant Me^{\alpha t}$ for some α) and has rational Laplace transform. Here we generalise this result to certain infinite dimensional systems.

5.2 Controllability and Observability

We have seen in the introduction that there is a 'duality' between controllability and observability in finite dimensions; namely, the system (5.1) is controllable if and only if the dual system

$$\dot{z} = A^T z + C^T v$$

$$w = B^T z$$

is observable. This same duality for infinite dimensional systems was only realised more recently, and was stated formally by Dolecki and Russell (1977) who showed that the concepts of controllability and observability can be reduced to certain duality theorems of functional analysis. We shall follow, essentially, their approach here, and begin with (what to the author) seems the more intuitive concept of controllability. Consider then the system

$$\begin{aligned} \dot{x} &= Ax + Bu \\ y &= Cx \end{aligned} \tag{5.3}$$

where A generates a strongly continuous semigroup $T(t)$ on a reflexive Banach space X and $B:U \to X$ and $C:X \to V$ are assumed to be bounded. (In the case of boundary control and observation, the same ideas apply; however, in this case, for example, $B:U \to Y$ for some Banach space Y and we have to assume that $T(t):Y \to X$ for $t > 0$.) We can write (5.3) in the 'mild' form

$$x(t) = T(t)x_0 + \int_0^\tau T(t-s)Bu(s)\mathrm{d}s \tag{5.4}$$

If U is the Banach space of controls then we consider the operator $\mathcal{B}:L^p[0,\tau;U] \to X$ defined by

$$\mathcal{B}u = \int_0^\tau T(t-s)Bu(s)\mathrm{d}s \qquad u \in L^p[0,\tau;U]$$

Suppose that we wish to drive any state x_0 to any other state x_1 in time τ with a control $u \in L^p[0,\tau;U]$. Then, since $T(\tau)x_0$ is fixed, we clearly require that

$$\mathcal{R}(\mathcal{B})\ (=\text{Range}\ (\mathcal{B})) = X \tag{5.5}$$

The condition (5.5) is very strong, however, and so we sometimes only require to get arbitrarily close to x_1, in which case we must have that

$$\overline{\mathcal{R}(\mathcal{B})} = X \tag{5.6}$$

If the system (5.3) (or (5.4)) is such that (5.5) holds, then we say that it is *exactly controllable* and if (5.6) holds, the system is called *approximately (or completely) controllable*. Exact controllability is a very strong condition, as mentioned above; in fact, it can be shown (Dolecki and Russell, 1977) that if the system (5.4) is exactly controllable, then the semigroup $T(t)$ can be extended to a group. We therefore will never expect to have exact controllability in the semigroup case, which applies, for example, to parabolic systems.

The range conditions above can be transformed into kernel conditions (which will make the duality between controllability and observability much clearer) by the following result.

Theorem 5.2.1

Let X_1, X_2, X_3 be Banach spaces and suppose that $F:X_1 \to X_3$, $G:X_2 \to X_3$ are bounded linear operators. Then,

$$\ker(G^*) \subseteq \ker(F^*) \leftrightarrow \overline{\text{Range}(G)} \supseteq \overline{\text{Range}(F)}$$

Proof: Suppose that $\ker G^* \subseteq \ker F^*$ and let $x \in \overline{\text{Range } F}$. If $x \notin \overline{\text{Range } G}$ then, by the Hahn-Banach theorem, there exists $x^* \in X_3^*$ *such that* $\langle x^*,x\rangle \neq 0$ but $\langle x^*,y\rangle = 0$ for all $y \in$ Range G. Hence,

$$\langle x^*,Gx_2\rangle = 0 \text{ for all } x_2 \in X_2$$

or

$$\langle G^*x^*,x_2\rangle_{X^*_2,X_2} = 0 \text{ for all } x_2 \in X_2$$

which means that $x^* \in \ker G^*$. But then $x^* \in \ker F^*$ and so

$$\langle x^*,Fx_1\rangle = 0 \text{ for all } x_1 \in X_1$$

However, since $x \in \overline{\text{Range } F}$ it follows that $\langle x^*,x\rangle = 0$, which is a contradiction. The converse implication is proved in a similar way. □

Corollary 5.2.2
The system (5.3) is approximately controllable iff the dual of the operator $\mathscr{B}$ defined above has trivial kernel. □

The following result (Curtain and Pritchard, 1978) is useful when considering exact controllability and can be proved quite easily.

Theorem 5.2.3
If U and X are reflexive Banach spaces (so that the dual semigroup $T^*(t)$ is strongly continuous on X^*), then the system (5.3) is (exactly) controllable iff there exists a $\gamma > 0$ such that

$$\gamma\|B^*T^*(.)x^*\|_{L^q[0,\tau;U^*]} \geqslant \|x^*\|_{X^*} \tag{5.7}$$

where

$$\frac{1}{p} + \frac{1}{q} = 1. \quad \square$$

Let us now consider observability of the system

$$\dot{x} = Ax \qquad x(0) = x_0$$
$$y = Cx \tag{5.8}$$

where $C:X \to V$ and V is the observation space. The system (5.8) is observable if, given the measurement or observation $y(t)$ for all $t \in [0,\tau]$, then one can determine (continuously) the initial state x_0. To make this statement more precise, let $Y = L^q[0,\tau;V]$, and note that

$$y(t) = CT(t)x_0$$

Define the operator $\mathscr{C}:X \to L^q[0,\tau;V]$ $(=Y)$ by

$$\mathscr{C}x = CT(.)x$$

Then $\mathscr{C}$ is called the observation operator and observability is equivalent to the existence of a bound k such that

$$\|x\|_X \leq k\|\mathscr{C}x\|_Y, \quad x \in \mathscr{D}(\mathscr{C}) \tag{5.9}$$

We emphasise the duality between controllability and observability now by evaluating the dual of the operator $\mathscr{C}$. Since the dual of $L^q[0,\tau;V]$ is just $L^p[0,\tau;V^*]$, $1/p + 1/q = 1$, we have, for any $f \in L^p[0,\tau;V^*]$:

$$\begin{aligned} \langle f(.), CT(.)x\rangle_{L^p[0,\tau;V^*],L^q[0,\tau;V]} &= \int_0^\tau \langle f(t),CT(t)x\rangle_{V^*V}\mathrm{d}t \\ &= \int_0^\tau \langle T^*(t)C^*f(t),x\rangle_{V^*,V}\mathrm{d}t \\ &= \langle \int_0^\tau T^*(t)C^*f(t)\mathrm{d}t,x\rangle_{V^*,V} \end{aligned}$$

and so

$$\begin{aligned} \mathscr{C}^*f &= \int_0^\tau T^*(t)C^*f(t)\mathrm{d}t \\ &= \int_0^\tau T^*(\tau - t)C^*h(t)\mathrm{d}t \end{aligned}$$

where $h(t) = f(\tau - t)$.

It follows, therefore, that the system (5.8) is observable iff the dual system

$$\dot{x}^* = A^*x^* + C^*u, \quad u \in V^*, \quad x^* \in X^*$$

is exactly controllable.

Just as in the controllability case, continuous observability is a strong condition and again (if C is bounded) implies that $T(t)$ can be extended to a group. For this reason we introduce F-observability as follows.

Definition 5.2.4

The system (5.8) is F-observable, where $F:X \to Z$ for some Banach space Z, if instead of (5.9) we have

$$\|Fx\|_Z < k\,\|\mathscr{C}x\|_Y \qquad x \in \mathscr{D}(\mathscr{C})$$

Of course, we can similarly generalise the concept of controllability.

Definition 5.2.5

The system (5.4) is G-controllable, where $G:W \to X^*$ for some Banach space W, if

$$\mathscr{R}(G) \subseteq \mathscr{R}(\mathscr{B})$$

Then we have the following theorem (Dolecki and Russell, 1977).

Theorem 5.2.6

The system (5.8) is F-observable iff the system

is F^*-controllable. □ (Note that since X is reflexive, A^* is the generator of the strongly continuous dual semigroup $T^*(t)$ of $T(t)$.) If the space W in definition 5.2.5 is dense in X^*, then we just have approximate controllability. We can use corollary 5.2.2 to prove the following generalisation of the finite dimensional rank condition stated in Section 5.1.

Lemma 5.2.7
Define

$$U_\infty = \{u \in U: Bu \in \mathscr{D}_\infty(A) \triangleq \bigcap_{n=1}^{\infty} \mathscr{D}(A^n)\}$$

Then the system (5.3) is approximately controllable on $[0,\tau]$ if

$$S \triangleq \overline{sp}\,[\cup\{A^n BU_\infty : n = 0,1,\ldots\}] = X$$

($\overline{sp}$ denotes the closed linear span of a set of vectors in X).

Proof: It follows from corollary 5.2.2 that (5.3) is approximately controllable iff

$$B^* T^*(t) x^* = 0 \text{ on } 0 \leqslant t \leqslant \tau$$

implies $x^* = 0$, and so if (5.3) is not approximately controllable on $[0,\tau]$ there exists $x^* \neq 0$ such that, for all $u \in U_\infty$,

$$\langle x^*, T(t)BU_\infty \rangle_{X^*,X} = 0 \text{ for } t \in [0,\tau]$$

Differentiating this countably many times at $t = 0$ gives

$$\langle x^*, A^n BU_\infty \rangle_{X^*,X} = 0,\ n = 0,1,2,\ldots$$

and so $S \neq X$. □

In the case when the control is a scalar, then $B = b$ is a vector in X and if $b \in \mathscr{D}_\infty(A)$ it follows that the system (5.3) is then approximately controllable if we have

$$\overline{sp}\{b, Ab, A^2b, \ldots\} = X \tag{5.10}$$

A vector b satisfying (5.10) is called a *cyclic vector* for A, and is important in realisation theory, as we shall see later. Note also that the converse of lemma 5.2.7 is not true (see Curtain and Pritchard, 1978).

We now give three examples of controllability which will show the application of the above methods. Since observability is dual to controllability, these examples also provide three cases of observability merely by taking the duals of each system.

Example 5.2.8
Consider the one-dimensional heat flow problem with pointwise control

$$\frac{\partial \phi}{\partial t} = \frac{\partial^2 \phi}{\partial x^2} + \delta(x - x_1)u \tag{5.11}$$

with $\phi(0,t) = \phi(1,t) = 0$. The semigroup generated by $A = d^2/dx^2$ with these boundary conditions is

$$T(t)\phi = \sqrt{2}\sum_{n=1}^{\infty}\phi_n e^{-n^2\pi^2 t}\sin(n\pi x),\ \phi_n = \sqrt{2}\langle \phi, \sin(n\pi x)\rangle$$

(*cf*. example 3.5.1). Now $U = \mathcal{R}$ and

$$B:U \to H^{-1/2-\varepsilon}([0,1])$$

for any $\varepsilon > 0$. Also, $B^*:H^{1/2+\varepsilon}([0,1]) \to \mathcal{R}$ is given by

$$B^*f = f(x_1),\ f \in H^{1/2+\varepsilon}([0,1])$$

and so it is easy to see that

$$B^*T^*(t) = \sqrt{2}\sum_{n=1}^{\infty}\phi_n e^{-n^2\pi^2 t}\sin(n\pi x_1)$$

and by corollary 5.2.2, the system (5.11) is approximately controllable iff

$$\sqrt{2}\sum_{n=1}^{\infty}\phi_n e^{-n^2\pi^2 t}\sin(n\pi x_1) = 0,\ \forall t \to \phi_n = 0,\ \forall n$$

By projecting into finite dimensional subspaces, it is easy to see that we must require x_1 to be an irrational number.

Example 5.2.9
We now consider the multipass process, defined in example 3.5.6, which takes the form

$$\frac{dx_k}{dt}(t) = A_0x_k(t) + A_1x_{k-1}(t) + \ldots + A_lx_{k-l}(t) + Bu_k,\ k \geqslant 0,\ t \in [0,\tau] \quad (5.12)$$

We have shown previously how to associate an operator $\mathcal{A}_m$, on a Hilbert space

$$\mathcal{H}_m = \bigoplus_{i=1}^{m} H$$

which generates a semigroup $\mathcal{T}_m(t)$ specified in example 3.5.6. $\mathcal{T}_m(t)$ can be represented in terms of an $(m + 1) \times (m + 1)$ matrix of operators in H in an obvious way and we denote the (i,j)th element of this matrix by $(\mathcal{T}_m(t))_{ij}$. Now define

$$G(u) \bigoplus_{i=1}^{m} L^2(0,\tau;U) \to L^2(0,\tau;H)$$

by

$$G(u) = \int_0^t \sum_{i=0}^{m-1}(\mathcal{T}_m(t-s))_{m,i+1}Bu_i(s)ds,\ 0 \leqslant t \leqslant \tau$$

We say that the system (5.12) is *approximately controllable* in m passes if it can be driven from any initial states $x_{-1},\ldots,x_{-l}$ to a dense subspace of $L^2([0,\tau];H)$.

To determine conditions for approximate controllability, we must find the dual of G. In fact,

$$\left\langle \int_0^{\cdot} \sum_{i=0}^{m-1} (\mathcal{T}_m(\cdot - s))_{m,i+1} Bu_i(s)\mathrm{d}s, h(\cdot)\right\rangle_{L^2(0,\tau;H),L^2(0,\tau;H)}$$

$$= \int_0^\tau \left\langle \int_0^t \sum_{i=0}^{m-1} (\mathcal{T}_m(t-s))_{m,i+1} Bu_i(s)\mathrm{d}s, h(t)\right\rangle_H \mathrm{d}t$$

$$= \sum_{i=0}^{m-1} \int_0^\tau \int_s^\tau \langle u_i(s), B^*(\mathcal{T}_m^*(t-s))_{i+1,m} h(t)\rangle_{U,U^*} \mathrm{d}t\mathrm{d}s$$

$$= \sum_{i=0}^{m-1} \left\langle u_i(\cdot), \int_{\cdot}^\tau B^*(\mathcal{T}_m^*(t-\cdot))_{i+1,m} h(t)\mathrm{d}t\right\rangle_{L^2(0,\tau;U),L^2(0,\tau;U^*)}$$

and so

$$G^*: L^2(0,\tau;H) \to \bigoplus_{i=1}^m L^2(0,\tau;U)$$

is given by

$$(G^*h)(s) = \left(\int_s^\tau B^*(\mathcal{T}_m^*(t-s))_{i+1,m} h(t)\mathrm{d}t\right)_{1\le i\le m}$$

It follows that (5.12) is approximately controllable in m passes iff

$$\int_s^\tau B^*(\mathcal{T}_m^*(t-s))_{i+1,m} h(t)\mathrm{d}t = 0, h \in L^2(0,\tau;H)$$

for almost all $s \in [0,\tau]$ and $0 \le i \le m-1$ implies

$$h(t) = 0 \text{ for almost all } t \in [0,\tau]$$

For more details on multipass systems see Banks (1982)

Example 5.2.10
Consider the system

$$\dot{x} = Ax + bu, x \in l^2 \tag{5.13}$$

where A is the right-shift operator, which has matrix representation

$$A = \begin{pmatrix} 0 & & & \\ 1 & 0 & & \underline{0} \\ & 1 & \cdot & \\ & \underline{0} & \cdot & \cdot & \cdot \end{pmatrix}$$

in the standard basis $\{e_i\}$ of l^2, and $b = (1,0,0,0,\ldots)^T$. Then

$$b = \begin{pmatrix} 1 \\ 0 \\ 0 \\ \vdots \end{pmatrix} = e_1, \quad Ab = \begin{pmatrix} 0 \\ 1 \\ 0 \\ \vdots \end{pmatrix} = e_2, \quad A^2b = \begin{pmatrix} 0 \\ 0 \\ 1 \\ 0 \\ \vdots \end{pmatrix} = e_3, \ldots$$

It follows that b is a cyclic vector for A and so (5.13) is approximately controllable.

In cases where the operator A of a system has compact resolvent, which implies that A has only a discrete point spectrum, it is often possible to reduce the controllability problem to an equivalent moment problem. We shall illustrate this with the following simple example, which is a special case of a system considered by Fattorini and Russell (1971). Let the state $y(x,t)$, $x \in [0,1]$, satisfy the linear parabolic equation

$$\begin{aligned} \frac{\partial y}{\partial t} &= \frac{\partial}{\partial x}\left(p(x)\frac{\partial y}{\partial x}\right) + q(x)y \\ y(0,t) &= g(t) \\ y(1,t) &= 0 \end{aligned} \tag{5.14}$$

Then, if $y(x,0) \in L^2([0,1])$, this system has a unique solution for any control g such that $g' \in L^2([0,1])$. Define the operator A by

$$(Ay)(x) = (p(x)y'(x))' + q(x)y(x)$$

with domain

$$\mathscr{D}(A) = \{y \in L^2([0,1]) : y', y'' \in L^2(0,1), y(0,t) = y(1,t) = 0\}$$

Then A has a sequence of a distinct real eigenvalues

$$\lambda_1 < \ldots < \lambda_n < \ldots, \lambda_n \to \infty$$

Let $\{\phi_n\}$ be the corresponding sequence of normalised eigenfunctions which form a basis of $L^2(0,1)$. Then we can write

$$y(x,t) = \sum_{n=1}^{\infty} \eta_n(t)\phi_n(x) \tag{5.15}$$

and

$$\begin{aligned} y_0(x) &\triangleq y(x,0) = \sum_{n=1}^{\infty} \mu_n\phi_n(x) \\ y_T(x) &\triangleq y(x,T) = \sum_{n=1}^{\infty} \upsilon_n\phi_n(x) \end{aligned}$$

for some fixed $T > 0$. The control g will therefore drive y_0 to y_T in time T if

$$\eta_n(T) = \upsilon_n, n = 1,2,\ldots \tag{5.16}$$

Consider now the adjoint system defined by

$$\frac{\partial w}{\partial t} = -\frac{\partial}{\partial x}\left(p(x)\frac{\partial w}{\partial x}\right) - q(x)w$$

$$w(0,t) = w(1,t) = 0$$

Let $D = \{(x,t): 0 \leqslant x \leqslant 1, 0 \leqslant t \leqslant T\}$. Then, by the divergence theorem,

$$0 = \int\int_D \left\{ w(x,t)\left[\frac{\partial y(x,t)}{\partial t} - \frac{\partial}{\partial x}\left(p(x)\frac{\partial y(x,t)}{\partial x}\right) - q(x)y(x,t)\right]\right.$$

$$\left. + y(x,t)\left[\frac{\partial w(x,t)}{\partial t} + \frac{\partial}{\partial x}\left(p(x)\frac{\partial w(x,t)}{\partial x}\right) + q(x)w(x,t)\right]\right\} \mathrm{d}x\mathrm{d}t$$

$$= \int\int_D \left\{\frac{\partial}{\partial x}\left[p(x)\left(y(x,t)\frac{\partial w(x,t)}{\partial x} - w(x,t)\frac{\partial y(x,t)}{\partial x}\right)\right]\right.$$

$$\left. + \frac{\partial}{\partial t}[y(x,t)w(x,t)]\right\} \mathrm{d}x\mathrm{d}t$$

$$= \int_0^1 [y(x,T)w(x,T) - y(x,0)w(x,0)]\mathrm{d}x$$

$$+ \int_0^T \left[p(1)\left(y(1,t)\frac{\partial w}{\partial x}(1,t) - w(1,t)\frac{\partial y}{\partial x}(1,t)\right)\right.$$

$$\left. -p(0)\left(y(0,t)\frac{\partial w}{\partial x}(0,t) - w(0,t)\frac{\partial y}{\partial x}(0,t)\right)\right] \mathrm{d}t \qquad (5.18)$$

Now, the function $w(x,t) = \phi_n(x)e^{\lambda_n(t-T)}$ is a solution of (5.17) for each $n \geqslant 1$ and so substituting this function for w and (5.15) for y in (5.18) we obtain

$$\eta_n(T) - e^{-\lambda_n T}\mu_n = \int_0^T e^{-\lambda_n(T-t)}p(0)\phi_n'(0)g(t)\mathrm{d}t$$

Hence by (5.16) we require

$$\upsilon_n - e^{-\lambda_n T}\mu_n = \int_0^T e^{-\lambda_n(T-t)}p(0)\phi_n'(0)g(t)\mathrm{d}t$$

Therefore, given υ_n,μ_n and T, we require to solve the *moment problem*

$$\int_0^T e^{-\lambda_n t}h(t)\mathrm{d}t = c_n \qquad (5.19)$$

where

$$c_n = (\upsilon_n - e^{-\lambda_n T}\mu_n)/(e^{-\lambda_n T}p(0)\phi_n'(0))$$

and $h(t) = g(T - t)$.

Define $L = \int_0^1 1/\sqrt{p(x)}\mathrm{d}x$. Then it can be shown that the moment problem (5.19) is soluble (i.e. y_0 can be driven to y_T) if

$$|\upsilon_n - \mathrm{e}^{-\lambda_n T}\mu_n| \leqslant M \exp[-(L+\eta)n],\ n \geqslant 1$$

for some $M,\eta > 0$.

Further results on controllability by moment sequences can be found in the literature (for example, Fattorini, 1975a). We note finally in this section that the controllability of nonlinear distributed systems has been considered by several authors, usually by applying the implicit function theorem (Fattorini, 1975*b*) to obtain local controllability or by obtaining specific solutions in the case of hyperbolic systems (Cirina, 1969). The reader should also consult the review paper of Russell (1978*a*) for a more complete discussion of controllability and observability.

5.3 Stabilisability

In many systems applications, we do not want to control from any given state to any other; we are often only interested in obtaining a feedback control which produces a stable system. In the case of the system

$$\dot{x} = Ax + Bu \tag{5.20}$$

where the state x belongs to a Banach space X and A generates a strongly continuous semigroup $T(t)$ on X, we wish to find a feedback $u = Dx$, where $D:\mathscr{D}(D) \subseteq X \to X$ such that $F = A + BD$ also generates a semigroup $T_F(t)$ for which

$$\|T_F(t)x\| \to 0 \text{ as } t \to \infty,\ x \in \mathscr{D}(A)$$

In such a situation we say that the system (5.20) is *stabilisable*. If we have, in addition,

$$\|T_F(t)x\| \leqslant C\mathrm{e}^{-at}\|x\|,\ t \geqslant 0$$

for some constants C,a, the system (5.20) is *exponentially stabilisable*. (Equivalently, we speak of the pair (A,B) as being stabilisable.)

In the introduction, we have seen the intuitively appealing result that controllability implies stabilisability and that stabilisability implies controllability of the unstable modes. The corresponding results for infinite dimensional systems are no longer true (indeed in the latter case we can only consider modal expansions for very special types of operators). The next example shows that approximate controllability does not imply stabilisability.

Example 5.3.1

Let A denote the right-shift operator on l^2 (*cf.* example 5.2.10), and let $b =$

$(1,0,0,\ldots)^T$. Then (A,b) is approximately controllable by example 5.2.10, but, for any $d = (d_1,d_2,\ldots) \in l^2$,

$$bd = \begin{pmatrix} d_1 & d_2 & \cdots \\ & \underline{0} & \end{pmatrix}$$

and $A + bd$ clearly has points λ in its spectrum for which $\mathrm{Re}\,\lambda > 0$. However, in Chapter 4 we showed that the stability of bounded operators is determined by the spectrum and so $A + bd$ is never stable for any $d \in l^2$. (This can also be seen by invoking results on the perturbation of the spectrum of A by the (compact) operator bd (Kato, 1976).)

It is clear from this example that for distributed systems more restrictive conditions than just approximate controllability are required for stabilisability. In fact, the following results, due to Slemrod (1974; 1976) can be proved.

Theorem 5.3.2
Let A generate a strongly continuous group $T(t)$ on a Hilbert space H such that for some $\varepsilon > 0$, there exists $\delta > 0$ such that

$$\int_0^{\varepsilon} \|B^*T^*(-t)y\|_U^2 \mathrm{d}t \geqslant \delta \|y\|_H^2, \ \forall y \in H$$

(U is also a Hilbert space and we have identified U with U^*). Then the pair (A,B) is stabilisable. Moreover, for any $\lambda > 0$, there exists $K \in \mathscr{L}(H,U)$ such that $A + BK$ is exponentially stable of order λ; i.e.

$$\|S(t)\|_{\mathscr{L}(H)} \leqslant M\mathrm{e}^{-\lambda t}$$

where S is generated by $A + BK$. □

Theorem 5.3.3
Let A generate a contraction group $T(t)$ on H (i.e. $\|T(t)\| \leqslant 1$, $\forall t$), and suppose that:

(a) there exists λ_0 such that, for $\lambda > \lambda_0$, $\lambda \in \rho(A)$ and $(A + \lambda I)^{-1}$ is compact;
(b) (A,B) is approximately controllable;
(c) $A = -A^*$.

Then (A,B) is stabilisable. □

Another approach to stabilisability, due to Triggiani (1975), is to consider the operator A as a direct sum $A_s \oplus A_u$, where s and u refer to the 'stable' and 'unstable' parts of A, respectively. More precisely, if we have the system

$$\dot{x} = Ax + Bu \tag{5.21}$$

then we suppose that the spectrum of A splits into two parts

$$\sigma(A) = \sigma_u(A) \cup \sigma_s(A)$$

so that a finite number of Jordan curves Γ_i $(1 \leqslant i \leqslant n)$ contains $\sigma_u(A)$ in the interior and excludes $\sigma_s(A)$ ($\sigma_u(A)$ is therefore bounded). As in general spectral theory (*cf.* Chapter 2), we define the projection

$$P = \frac{1}{2\pi i} \sum_{j=1}^{n} \int_{\Gamma_j} R(\lambda;A)^{-1} d\lambda$$

Then P is bounded and we write

$$x_u = Px, \; x_s = (I - P)x$$

so that (5.21) becomes

$$\dot{x}_u = A_u x_u + PBu \tag{5.22a}$$

$$\dot{x} = A_s x_s + (I - P)Bu \tag{5.22b}$$

where

$$A_u = A\big|_{PX}, \qquad A_s = A\big|_{(I-P)X}$$

Suppose now that $\sigma(A_s)$ is contained in $\{\lambda:\text{Re}(\lambda) \leqslant -\delta\}$ for some $\delta > 0$ and that A_s satisfies the spectrum determined growth condition (*cf.* Chapter 4) on $X_s = (I - P)X$. Then it is easy to show that if (5.22*a*) is exponentially stabilisable using the feedback control $u = D_u x_u$, the original system (5.21) is also exponentially stabilisable using the feedback control $D = (D_u, 0): X_u \oplus X_s \to U$.

The assumption that only a finite number of eigenvalues lie in the right-half plane can be removed, in the case of the heat equation, as shown by Feintuch and Rosenfeld (1978), and also by Hamatsuka *et al.* (1981).

Before closing this section we remark that, with reference to the spectral assignability problem, Russell (1978*b*) has proved the following result by reducing the system to canonical forms similar to those in the finite dimensional case.

Theorem 5.3.4

Consider the controlled hyperbolic system

$$\frac{\partial}{\partial t}\begin{pmatrix} w \\ v \end{pmatrix} - \begin{pmatrix} 0 & 1 \\ 1 & 0 \end{pmatrix} \frac{\partial}{\partial x}\begin{pmatrix} w \\ v \end{pmatrix} - A(x)\begin{pmatrix} w \\ v \end{pmatrix} = g(x)u(t) \tag{5.23}$$

where $A(x)$ is a continuous matrix and $g \in L^2([0,1];\mathcal{R}^2)$, together with the boundary conditions

$$a_0 w(0,t) + b_0 v(0,t) = 0, \; a_0 \neq 0 \text{ or } b_0 \neq 0$$

$$a_1 w(1,t) + b_1 v(1,t) = 0, \; a_1 \neq 0 \text{ or } b_1 \neq 0$$

Let σ_k $(k = 0, \pm 1, \pm 2, \ldots)$ denote the eigenvalues of the operator

$$L_0 : \begin{pmatrix} 0 & 1 \\ 1 & 0 \end{pmatrix} \frac{\partial}{\partial x} + A(x)$$

and let

$$\Phi_k = \begin{pmatrix} \phi_k^1(x) \\ \Phi_k^2(x) \end{pmatrix}$$

denote the corresponding eigenfunctions. If

$$g = \sum_{k=-\infty}^{\infty} g_k \Phi_k$$

and $\{\rho_j\}$ is a sequence of numbers satisfying

$$\sum_{j=-\infty}^{\infty} \left| \frac{\rho_j - \sigma_j}{g_j} \right|^2 < \infty$$

then there exists $k \in L^2([0,1];\mathscr{R}^2)$ such that the system (5.23) with the control

$$u = \left(\begin{pmatrix} w(.,t) \\ v(.,t) \end{pmatrix}, k \right)_{L^2([0,1];\mathscr{R}^2)}$$

has eigenvalues ρ_j, $j = 0, \pm 1, \pm 2, \ldots$ □

5.4 Realisability

A finite dimensional linear single-input-single-output system is specified, in the state space formulation, by the triple (A,b,c) consisting of a matrix A and vectors b,c (of appropriate size). This triple gives rise to a transfer function

$$G(s) = c(sI - A)^{-1}b$$

and it is well known that if (A,b,c) is controllable and observable, then the poles of the rational function G correspond (counting multiplicities) to the spectral values of A. We then say that the representation (A,b,c) of G is *minimal*. Similar remarks apply to the multivariable case (A,B,C) except that $G(s) = C(sI - A)^{-1}B$ may have poles which have lower multiplicity than the corresponding spectral values of A. As we shall see, the situation is not quite so easy in the case of distributed systems, and so we introduce the following definitions (the material in this section is mainly from Baras and Brockett, 1975).

Definition 5.4.1
Let $g:[0,\infty) \to \mathscr{R}$ be a given continuous function, and let H be a Hilbert space. Then we say that g *has the (balanced) realisation* (A,b,c) on H, where A is a

closed operator which generates a semigroup $T(t)$ and $b \in \mathscr{D}(A)$, $c \in \mathscr{D}(A)^*$ ($\mathscr{D}(A)$ has the graph norm), if

$$g(t) = \langle c, T(t)b \rangle \tag{5.24}$$

If $b,c \in H$, then the realisation is said to be *regular*, and if A is bounded then we say that the realisation is *bounded*. In the latter case,

$$g(t) = \langle c, e^{At}b \rangle \tag{5.25}$$

Definition 5.4.2
A realisation (A,b,c) of g is called *canonical* if b is a cyclic vector for A and c is a cyclic vector for A^*. It is called *S-minimal* if $\sigma(A) = \tilde{\sigma}(G) = \mathscr{L}(g)(s)$ (the Laplace transform of g) and

$$\tilde{\sigma}(G) = \{s \in \mathscr{C}: G \text{ is not analytic at } s\}$$

Theorem 5.4.3
The function $g:[0,\infty) \to \mathscr{R}$ has a bounded realisation iff g is an entire function of exponential order (i.e. $|g(t)| \leqslant Me^{at}$ for some M,a).

Proof: The necessity is obvious. For sufficiency note that g is entire and so has an everywhere convergent Taylor series

$$g(t) = \sum_{n=0}^{\infty} g_n t^n$$

If σ_0 is the exponential order of g, then $\overline{\lim}_{n\to\infty} (n!|g_n|)^{1/n} = \sigma_0$ and so, for $k > \sigma_0$, the sequence $\{n!|g_n|/k^n\}_{n\geqslant 0} \in l^2$. We can therefore choose $H = l^2$, and

$$A = k \begin{bmatrix} 0 & & & \\ 1 & 0 & 0 & \\ & 1 & \bar{0} & \\ & \underline{0} & & \cdots \\ & & & \cdots \end{bmatrix}$$

$$b = \{1,0,0,\ldots.\}$$

$$c = \{g_0, g_1/k, \ldots, n\, g_n/k^n, \ldots\}. \quad \square$$

In a similar way, we can show that $g:[0,\infty) \to \mathscr{R}$ has a bounded realisation iff $G(s) = \mathscr{L}(g)(s)$ is analytic at infinity and vanishes there. Indeed, the necessity is again obvious and for sufficiency we note that

$$G(s) = \sum_{i=0}^{\infty} a_i s^{-(i+1)} \text{ for } |s| > c \text{ (large } c\text{)}$$

Then for $k > c$ we take A to be k times the right shift as in (5.26) and $b = \{1,0,0,\ldots\}$, $c = \{a_0,a_1/k,a_2/k^2,\ldots\}$.

Note that in the above realisations we have chosen k so that all the singularities of G are contained in the disc of radius k (which is the spectrum of kA) and so they may be far from being S-minimal. In order to obtain canonical realisations, however, we can use the following result.

Theorem 5.4.4
Let $g(t) = \langle c,e^{At}b\rangle$, and define

$$M = \overline{sp}\{c,A^*c,A^{*2}c,\ldots\}$$

$$N = \overline{sp}\{P_Mb,\ldots,(P_MAP_M)^iP_Mb,\ldots\}$$

where P_m is the projection on M. Define P_N similarly. Then

$$g(t) = \langle P_Nc,e^{(P_NAP_N)t}P_Mb\rangle. \quad \square$$

Now in the above realisations, A is (k times) the right shift operator and b is cyclic for A. Hence $N = M$ and so we can replace (A,b,c) by the canonical realisation (P_MAP_M,P_Mb,c) with state space M.

The general theory of S-minimal realisations remains unclear, although results in certain cases can be obtained (see, for example, Baras and Brockett, 1975 and Helton, 1976).

Finally in this chapter we discuss realisations which are not bounded. Note first that it is easy to see that $g(t)$ has a balanced realisation iff it has a regular one and so we can restrict attention to the former case. We then have the following theorem.

Theorem 5.4.5
If $g(t)$ is realisable (i.e. has a balanced realisation) then it is continuous and of exponential order. Conversely, if $g(t)$ is absolutely continuous on bounded intervals and $\dot{g}(t)$ is of exponential order (i.e. ess sup $|\dot{g}(t)| \leqslant Ke^{at}$ for some K,a), then $g(t)$ is realisable.

Proof: If g has a balanced realisation, it also has a regular one and so $g(t) = \langle c,T(t)b\rangle$, where T is a semigroup generated by some operator A. Then g is clearly continuous and by the Hille-Yosida theorem (theorem 3.2.6) we have $|g(t)| \leqslant \|c\|\cdot\|b\|Me^{\alpha t}$ for some M,α.

Conversely, if g satisfies the hypotheses of the theorem then for large σ, $e^{-\sigma t}\dot{g}(t) \in L^2(0,\infty)$. Hence the function $e^{-\sigma t}g(t)$, which also belongs to $L^2(0,\infty)$, is locally absolutely continuous and its derivative is in $L^2(0,\infty)$. Consider the operator $A = \partial/\partial x$ on $L^2(0,\infty)$. Then $\mathscr{D}(A)$is dense in $L^2(0,\infty)$ and consists of locally absolutely continuous functions whose derivative is in $L^2(0,\infty)$. If we define

$$(T(t)f)(x) = f(x + t),\ f \in L^2(0,\infty)$$

then $\dot{T}f = ATf, f \in \mathscr{D}(A)$ and so $T(t)$ is the semigroup generated by A (T is the translation semigroup, and corresponds to the shift operator for bounded realisations). Now let c be the functional on $\mathscr{D}(A)$ defined by

$$c(f) = f(0)$$

Now,

$$|c(f)|^2 = |f(0)|^2 \leqslant 2\int_0^\infty |f(x)|\cdot|f'(x)|\mathrm{d}x$$

$$< \int_0^\infty |f(x)|^2\mathrm{d}x + \int_0^\infty |f'(x)|^2\mathrm{d}x$$

and so $c \in \mathscr{D}(A)^*$. If we take $b = \mathrm{e}^{-\sigma x}g(x)$ ($\in \mathscr{D}(A)$), then

$$c(T(t)b) = c[\mathrm{e}^{-\sigma(t+x)}g(t + x)] = \mathrm{e}^{-\sigma t}g(t)$$

and so

$$g(t) = c(S(t)b)$$

where S is the semigroup generated by $A + \sigma I$. □

We note finally that the spectral minimality of realisations of normal symmetric systems has been studied by Brockett and Fuhrmann (1976), and the reader should consult this paper for further information and other references. The theory of bounded realisations will have important applications to the generalised root locus, as we shall see in the last chapter.

Chapter 6

Optimal control theory

6.1 Introduction

We come now to study the optimal control of distributed systems; however, before proceeding with the general theory we shall remind the reader of the finite dimensional results. The most important problem in finite dimensional optimal control theory (mainly because of its solubility) is that posed by a linear equation

$$\dot{x} = Ax + Bu \tag{6.1}$$

together with a quadratic cost functional

$$J(u) = \|x(t_f)\|_G^2 + \int_{t_0}^{t_f} \{\|x(s)\|_D^2 + \|u(s)\|_R^2\} \mathrm{d}s \tag{6.2}$$

where $\|y\|_E \triangleq \langle y, Ey \rangle = y^T Ey$, for some matrix E, and D, G and R are symmetric matrices which are non-negative (and R strictly positive, i.e. invertible). This is the so-called *linear-quadratic problem*; we seek to minimise (6.2) subject to the constraint (6.1). It is well known that the optimal control is given by

$$u_{\text{op}} = -R^{-1}B^T Q(t)x(t) \tag{6.3}$$

where Q is the (unique) solution of the Riccati equation

$$\dot{Q}(t) = -A^T Q(t) - Q(t)A - D + Q(t)BR^{-1}B^T Q(t)$$

with the final condition

$$Q(t_f) = G$$

Hence, the solution is a feedback control with time-dependent 'gain'.

One drawback with this control law is that the control responds to large or small errors in essentially the same way. However, it may be that the system is subject to noise perturbations which do not correspond to 'true' error signals and so one would like to find a (nonlinear) control law which responds quickly to large errors, but more slowly to small perturbations. To obtain such a control we use the *receding horizon solution*. This control is obtained by considering the system (6.1) subject to the constraint $x(t_f) = 0$ and minimising, instead of (6.2), the cost functional

$$J(u) = \int_{t_0}^{t_f} \|u(s)\|_R^2 \mathrm{d}s \tag{6.4}$$

(*cf*. Shaw, 1979). If the pair (A,B) is controllable, then the optimal control is

$$u_{\text{op}} = -R^{-1}B^T \mathrm{e}^{-A^T(t-t_0)} W^{-1}(t_f - t_0)x_0,\ t_0 \leqslant t \leqslant t_f \tag{6.5}$$

where W satisfies

$$\dot{W} = BR^{-1}B^T - AW - WA^T;\ W(0) = 0 \tag{6.5a}$$

The optimal control (6.5) is open loop, however, and so we now apply the receding horizon philosophy, which says that we apply (6.5) as if, at each time, we were beginning a new $(t_f - t_0)$-second minimisation; i.e. put $t_0 = t$, $x_0 = x(t)$. Then

$$u = -R^{-1}B^T W^{-1}(t_f - t_0)x(t) \tag{6.6}$$

Of course, u is no longer optimal, but we can make this control u nonlinear by allowing $t_f - t_0$ to depend on the state x. This can be done in a variety of ways (Shaw, 1979; Banks, 1980) and it is found that a control with the desired properties can be obtained.

The most general necessary conditions for the optimal control of nonlinear systems are given by the maximum principle of Pontryagin (Pontryagin *et al*., 1962) and a unified approach to such problems using the theory of cones can be found in Girsanov (1972). However, the majority of papers concerned with the control of distributed systems seem to consider only the above linear-quadratic and receding horizon solutions, and, to some extent, the time optimal problem. We shall therefore concentrate on these special cases in this chapter.

6.2 The linear-quadratic problem

We shall consider the linear system

$$\dot{x} = Ax + Bu,\ x(0) = x_0 \in H \tag{6.7}$$

where A generates a semigroup $T(t)$ on a Hilbert space H and B is a bounded operator from the control Hilbert space U into H (i.e. $B \in \mathscr{L}(U,H)$; the case of boundary controls will be mentioned later). However, because a strict solution of (6.7) belongs to $\mathscr{D}(A)$, it is simpler to consider the integrated form (mild solution)

$$x(t) = T(t)x_0 + \int_0^t T(t-s)Bu(s)\mathrm{d}s \tag{6.8}$$

The cost functional which we take is the quadratic form

$$J(u) = \langle Gx(t_f),x(t_f)\rangle_H + \int_0^{t_f} (\langle Dx(t),x(t)\rangle_H + \langle Ru(t),u(t)\rangle_U)\mathrm{d}t \qquad (6.9)$$

where $G,D \in \mathscr{L}(H,H)$ are self-adjoint and non-negative , $R \in \mathscr{L}(U,U)$ with $\langle Ru,u\rangle \geqslant m\|u\|^2$ for some $m > 0$.

The problem (6.8), (6.9) has been solved in many ways in the literature (for example, Lions, 1971; Curtain and Pritchard, 1978); however, the approach which is perhaps conceptually the clearest is due to Gibson (1979) and it is this method which we shall follow. First we introduce the spaces

$$\mathscr{H}_t = L^2(0,t;H),\ \mathscr{U}_t = L^2(0,t;U)$$

and define the operators $\mathscr{T} \in \mathscr{L}(\mathscr{H}_{t_f},\mathscr{H}_{t_f})$, $\mathscr{T}_{t_f} \in \mathscr{L}(\mathscr{H}_{t_f},H)$ by

$$(\mathscr{T}\phi)(t) = \int_0^t T(t-s)\phi(s)\mathrm{d}s,\ \forall\phi \in \mathscr{H}_{t_f},\ 0 \leqslant t \leqslant t_f$$

$$(\mathscr{T}_{t_f}\phi) = (\mathscr{T}\phi)(t_f) \qquad \forall\phi \in \mathscr{H}_{t_f}$$

Then we can write the cost functional (6.9) in the form

$$\begin{aligned} J(u) = {} & \langle G(T(t_f)x_0 + \mathscr{T}_{t_f}Bu),\ (\mathscr{T}(t_f)x_0 + \mathscr{T}_{t_f}Bu)\rangle_H \\ & + \langle D(T(.)x_0 + \mathscr{T}Bu),\ (T(.)x_0 + \mathscr{T}Bu)\rangle_{\mathscr{H}_{t_f}} \\ & + \langle Ru,u\rangle_{\mathscr{U}_{t_f}} \end{aligned} \qquad (6.10)$$

The minimum of this quadratic functional (which can be shown to be unique; Lions, 1971) is then given by the solution of

$$J'(u)v = 0,\ \forall v \in \mathscr{U}_{t_f} \qquad (6.11)$$

where $'$ denotes the Fréchet derivative (*cf*. Chapter 2). Now,

$$\begin{aligned} J'(u)v &= 2\langle G(T(t_f)x_0 + \mathscr{T}_{t_f}Bu),\ \mathscr{T}_{t_f}Bv\rangle_H \\ &\quad + 2\langle D(T(.)x_0 + \mathscr{T}Bu),\ \mathscr{T}Bv\rangle_{\mathscr{H}_{t_f}} + 2\langle Ru,v\rangle_{\mathscr{U}_{t_f}} \\ &= 2\langle Gx(t_f),\mathscr{T}_{t_f}Bv\rangle_H + 2\langle Dx,\ \mathscr{T}Bv\rangle_{\mathscr{H}_{t_f}} + 2\langle Ru,v\rangle_{\mathscr{U}_{t_f}} \\ &= 2\langle B^*(\mathscr{T}^*_{t_f}Gx(t_f) + \mathscr{T}^*Dx) + Ru,v\rangle_{\mathscr{U}_{t_f}} \end{aligned}$$

where the dual operators $\mathscr{T}^*$ and $\mathscr{T}^*_{t_f}$ are given by

$$(\mathscr{T}^*\phi)(t) = \int_t^{t_f} \mathscr{T}^*(s-t)\phi(s)\mathrm{d}s$$

and

$$(\mathscr{T}^*_{t_f} x)(t) = \mathscr{T}^*(t_f - t)x$$

Hence, by (6.11), the optimal control is given by

$$u_{\mathrm{op}} = -R^{-1}B^*(\mathscr{T}^*_{t_f}Gx(t_f) + \mathscr{T}^*Dx) \qquad (6.12)$$

Using the fact that any part of an optimal trajectory is optimal, we may write

$$x(t) = S(t,s)x(s)$$

for the solution of (6.8) with the feedback control (6.12), and it can be shown that $S(t,s)$ is an evolution operator. Hence, we may write (6.12) in the form

$$u_{op}(t) = -R^{-1}B^*Q(t)x(t), \text{ a.e.} \tag{6.12a}$$

where

$$Q(t)x = T^*(t_f - t)GS(t_f,t)x + \int_t^{t_f} T^*(s-t)DS(s,t)x\,ds,\ x \in H \tag{6.13}$$

However, S is the perturbation of $T(t)$ by $-BR^{-1}B^*Q(t)$ and so S satisfies the integral equation

$$S(t,s)x = T(t-s)x - \int_s^t S(t,\tau)BR^{-1}B^*Q(\tau)T(\tau - s)x\,d\tau,\ x \in H \tag{6.14}$$

$$\left(= T(t-s)x - \int_s^t T(t-\tau)BR^{-1}B^*Q(\tau)S(\tau,s)x\,d\tau\right)$$

and it follows that $Q(t)$ satisfies the integral Riccati equation

$$Q(t)x = T^*(t_f - t)\,\boldsymbol{G}\boldsymbol{T}(t_f - t)x + \int_t^{t_f} T^*(\tau - t)[D - Q(\tau)BR^{-1}B^*Q(\tau)]T(\tau - t)x\,d\tau$$

for $0 \leqslant t \leqslant t_f$ and $x \in H$. $Q(t)$ is therefore self-adjoint, and from (6.13) we obtain

$$\begin{aligned}\langle Q(t)x,y\rangle_H &= \langle GS(t_f,t)x, T(t_f - t)y\rangle_H + \int_t^{t_f} \langle DS(\tau,t)x, T(\tau - t)y\rangle_H d\tau \\ &= \langle GS(t_f,t)x, S(t_f,t)y\rangle_H \\ &\quad + \int_t^{t_f} \langle GS(t_f,t)x, T(t_f - \tau)BR^{-1}B^*Q(\tau)S(\tau - t)y\rangle_H d\tau \\ &\quad + \int_t^{t_f} \langle DS(\tau,t)x, S(\tau,t)y\rangle_H d\tau \\ &\quad + \int_t^{t_f} \langle DS(\tau,t)x, \int_t^{\tau} T(\tau - s)BR^{-1}B^*Q(s)S(s,t)y\,ds\rangle_H d\tau\end{aligned}$$

by (6.14). Hence, by simple manipulation,

$$\begin{aligned}\langle Q(t)x,y\rangle_H = \langle GS(t_f,t)x, S(t_f,t)y\rangle_H + \int_t^{t_f} (\langle DS(\tau,t)x, S(\tau,t)y\rangle_H \\ + \langle RR^{-1}B^*Q(\tau)S(\tau,t)x, R^{-1}B^*Q(\tau)S(\tau,t)y\rangle_U)d\tau\end{aligned} \tag{6.15}$$

and so $\langle Q(t)x,x\rangle_H$ is just the cost function (6.9) with u given by the feedback control (6.12); i.e.

$$J(u_{op}) = \langle Q(0)x,x\rangle_H \tag{6.16}$$

We also see that $Q(t) \geqslant 0$. From (6.15) we have

$$\langle Q(t)x,x\rangle_H = \langle S^*(t_f,t)GS(t_f,t)x,x\rangle_H$$

$$+ \langle \int_t^{t_f} S^*(\tau,t)[D + Q(\tau)BR^{-1}B^*Q(\tau)]S(\tau,t)x\mathrm{d}\tau,x\rangle_H$$

and so it follows that

$$Q(t)x = S^*(t_f,t)GS(t_f,t)x + \int_f^{t_f} S^*(\tau,t)[D + Q(\tau)BR^{-1}B^*Q(\tau)]S(\tau,t)x\mathrm{d}\tau$$

for $t \in [0,t_f]$. It can be shown that the solution $Q(t)$ of this integral equation is unique and satisfies the inner product Riccati equation

$$\frac{\mathrm{d}}{\mathrm{d}t}\langle Q(t)x,y\rangle + \langle Q(t)x,Ay\rangle + \langle Ax,Q(t)y\rangle + \langle Dx,y\rangle$$

$$= \langle Q(t)BR^{-1}B^*Q(t)x,y\rangle \quad \text{on } [0,t_f] \tag{6.17}$$

and

$$Q(t_f) = G$$

for $x,y \in \mathscr{D}(A)$ (see Curtain and Pritchard, 1978).

The above reasoning leads to the Riccati equation in various forms. We can also use essentially the same approach for boundary control (when B is not a bounded operator from U to H) by introducing the assumption:

$B \in \mathscr{L}(U,W)$, for a new Hilbert space W, such that

(*a*) $H \subseteq \mathscr{R}(B) \subseteq W$
(*b*) $T(t) \in \mathscr{L}(W,H), t > 0$
(*c*) $\|T(t)w\|_H \leqslant g(t)\|w\|_W, \ \forall w \in W$

where $g \in L^q[0,t_f]$.

Then the discussion above is formally the same, with the exception, however, that now $Q(t) \in \mathscr{L}(W,H)$ for almost all t. Using similar methods to those above, we can also consider the tracking problem; i.e. where the cost functional is now of the form

$$J(u) = \langle G(x(t_f) - r(t_f)),x(t_f) - r(t_f)\rangle_H$$

$$+ \int_0^{t_f} (\langle D(x(t) - r(t)),x(t) - r(t)\rangle_H + \langle Ru(t),u(t)\rangle_U)\mathrm{d}t$$

for some given function r. Then we obtain the optimal control

$$u_{\mathrm{op}}(t) = -R^{-1}B^*Q(t)x(t) - R^{-1}B^*s(t) \tag{6.18}$$

where the first term is as before and $s(t)$ is the unique solution of the equation

$$\frac{\mathrm{d}}{\mathrm{d}t}\langle s(t),x\rangle = -\langle s(t),(A - BR^{-1}B^*Q(t))x\rangle + \langle Dr(t),x\rangle$$

and

$$s(t_f) = Gr(t_f)$$

for $x \in \mathscr{D}(A)$.

Before giving some examples we note finally that if we consider the infinite time cost

$$J(u) = \int_0^\infty (\langle Dx(t),x(t)\rangle_H + \langle Ru(t),u(t)\rangle_U)\mathrm{d}t$$

then we obtain the optimal control

$$u_{\mathrm{op}}(t) = -R^{-1}B^*Qx(t)$$

where Q satisfies the algebraic Riccati equation

$$\langle Ax,Qy\rangle + \langle Qx,Ay\rangle + \langle Dx,y\rangle = \langle QBR^{-1}B^*Qx,y\rangle$$

provided $(A^*,D^{1/2})$ and (A,B) are stabilisable pairs (*cf.* Curtain and Pritchard, 1978).

6.3 Examples

In this section we shall give some examples of the above theory. The first two examples are well known and so the results will be stated without proof; these examples are included for comparison purposes when we consider the receding horizon solution.

Example 6.3.1
We consider the heat equation

$$x_t = x_{\xi\xi} + u(t,\xi), \qquad \xi \in [0,1] \tag{6.19}$$

$$x_\xi(0,t) = x_\xi(1,t) = 0$$

$$H = L^2[0,1]$$

together with the cost functional

$$J(u) = \int_0^1 x^2(t_f,\xi)\mathrm{d}\xi + \int_0^{t_f}\left(\int_0^1 \{x^2(t,\xi) + u^2(t,\xi)\}\mathrm{d}\xi\right)\mathrm{d}t$$

We can write

$$Q(t)h = \sum_{i=0}^{\infty}\sum_{j=0}^{\infty} q_{ij}(t)\phi_j\langle h,\phi_i\rangle, \; h \in L^2[0,1]$$

where $\phi_i = \sqrt{2}\cos \pi i\xi$, and we have

$$q_{ij}(t) \equiv 0, \; i \neq j$$

$$q_{ii}(t) = \frac{a_i(1-b_i) - b_i(1-a_i)\mathrm{e}^{-\alpha_i(t-t_f)}}{(1-b_i) - (1-a_i)\mathrm{e}^{-\alpha_i(t-t_f)}}$$

with

$$\alpha_i = 2(\pi^4 i^4 + 1)^{1/2}$$
$$a_i = -\pi^2 i^2 - \alpha_i/2$$
$$b_i = -\pi^2 i^2 + \alpha_i/2$$

The optimal control is

$$u_{op}(t) = \sum_{i=0}^{\infty} u_i(t)\phi_i$$

where

$$u_i(t) = -q_{ii}(t)x_i(t) \qquad (x_i(t) = \langle x,\phi_i\rangle)$$

Example 6.3.2
The controlled wave equation

$$x_{tt} = x_{\xi\xi} + u(t,\xi),\ \xi \in [0,1] \tag{6.21}$$
$$x(0,t) = x(1,t) = 0, x(\xi,0) = x_0(\xi), x_t(\xi,0) = x_1(\xi)$$
$$H = L^2[0,1]$$

and the cost functional

$$J(u) = \tfrac{1}{2}\int_0^1 (x^2(\xi,t_f) + x_t^2(\xi,t_f))\mathrm{d}\xi + \int_0^{t_f}\int_0^1 (\tfrac{1}{4}x_t^2(\xi,t) + u^2(\xi,t))\mathrm{d}\xi\mathrm{d}t$$

may be written in the forms

$$\boldsymbol{x}(t) = \begin{pmatrix} x(t) \\ x_t(t) \end{pmatrix} = T(t)\begin{pmatrix} x_0 \\ x_1 \end{pmatrix} + \int_0^t T(t-s)\begin{pmatrix} 0 \\ \mathrm{I} \end{pmatrix} u(s)\mathrm{d}s$$

where the group T is defined in example (3.5.3) and

$$J(u) = \tfrac{1}{2}\langle \boldsymbol{x}(t_f),\boldsymbol{x}(t_f)\rangle + \int_0^{t_1} (\tfrac{1}{4}\langle D\boldsymbol{x}(t),\boldsymbol{x}(t)\rangle_{\mathscr{H}} + \langle u,u\rangle_H)\mathrm{d}t$$

where

$$\mathscr{H} = \mathscr{D}((-A)^{1/2}) \oplus H, \quad D = \begin{pmatrix} 0 & 0 \\ 0 & \mathrm{I} \end{pmatrix}$$

and

$$\langle \boldsymbol{x},\boldsymbol{y}\rangle_{\mathscr{H}} = \langle x_{1\xi},y_{1\xi}\rangle_H + \langle x_2,y_2\rangle_H\ (\boldsymbol{x} = (x_1,x_2), \boldsymbol{y} = (y_1,y_2))$$

The optimal control is given by

$$u_{op} = -\,x_t(t,\xi)$$

Example 6.3.3
Consider the temperature control of a material body Ω by point actuators on the surface, as in Fig. 6.1. This is important in the control of residual stress in metals (*cf.* Kusakabe and Mihara, 1980, and Banks, 1981*c*). The actuators are assumed to be point controllers, but in fact will take effect over small areas a_i. However, it will be simpler to assign to such a controller a δ function of strength $T_i a_i$. Finally, the ambient temperature T_a is normalised to zero. Then it can be shown that the equation describing the controlled heat flow is

$$\frac{\partial T}{\partial t} = \mu \nabla^2 T + \mu H \sum_{i=1}^{N} u_i a_i \delta(p - p_i),\ u_i = T_i \qquad (6.22)$$

where μ is the thermal conductivity and H is the coefficient of heat transfer.

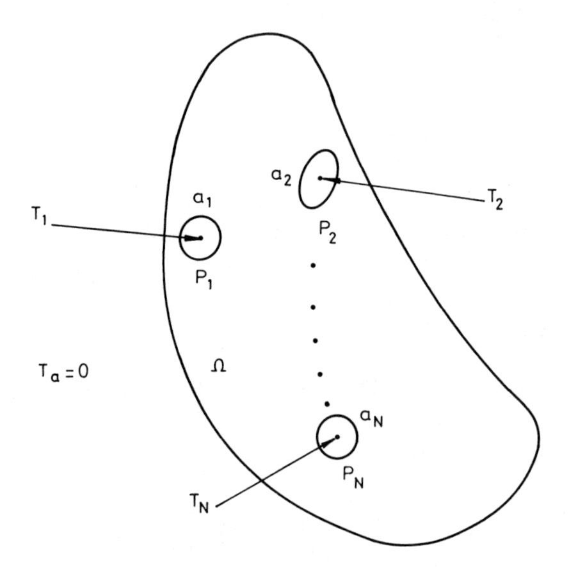

Fig. 6.1 *Material body with temperature controls T_i*

The boundary condition is

$$HT\bigg|_{\partial\Omega} = -\lambda \frac{\partial T}{\partial n}\bigg|_{\partial\Omega}$$

Note that if

$$\boldsymbol{B}\boldsymbol{u} = H \sum_{i=1}^{N} u_i a_i \delta(p - p_i),\ \boldsymbol{u} = (u_1,\ldots,u_N)$$

then

$$B \in \mathscr{L}(\mathscr{R}^N, H^{-3/2-\varepsilon}(\Omega))$$

It is easy to see that the semigroup generated by $\mu\nabla^2$ maps $H^{-3/2-\varepsilon}(\Omega)$ into $L^2(\Omega)$.

Suppose that we wish to minimise the variation in temperature in Ω from the mean value. This mean value is, of course,

$$T_m(t) = \frac{1}{V}\int_\Omega T(t,x)\mathrm{d}x, \; V \triangleq \int_\Omega \mathrm{d}x$$

It is then reasonable to minimise $\|T_v\|^2_{L^2(\Omega)}$, where

$$T_v = T(t,x) - T_m(t)$$

Define the averaging operator K on $L^2(\Omega)$ by

$$(K\phi)(x) = \frac{1}{V}\int_\Omega \phi(x')\mathrm{d}x'$$

(i.e. $K\phi$ is the constant function with the average of ϕ as its unique value). Then K is bounded and

$$\begin{aligned}\|T_v\|_{L^2(\Omega)} &= \langle (I-K)T,(I-K)T\rangle_{L^2(\Omega)} = \langle T,(I-K)^*(I-K)T\rangle_{L^2(\Omega)} \\ &= \langle T,(I-K)T\rangle_{L^2(\Omega)}\end{aligned}$$

since, as is easily seen, K is self-adjoint and idempotent (i.e. $K = K^2$). The cost functional can therefore be written

$$J(u) = \langle T(t_f),DT(t_f)\rangle_{L^2(\Omega)} + \int_0^{t_f} \{\langle T(s),DT(s)\rangle_{L^2(\Omega)} + \langle u,Ru\rangle_{\mathscr{R}^n}\}\,\mathrm{d}s$$

where $D = I - K$. Using the above results we see that the optimal control is given by

$$u_{op}(t) = -R^{-1}B^*Q(t)T(t,\xi)$$

where Q is given by the Riccati equation (6.17). Note that

$$B^*v = (\mu H a_i v(p_i))^T \in \mathscr{R}^N, \text{ for } v \in H^{3/2+\varepsilon}(\Omega)$$

and by Schwartz's kernel theorem (2.5.11), we may write

$$(Q(t)T)(x) = \int_\Omega \kappa(x,y,t)T(y)\mathrm{d}y$$

for some distribution κ. From (6.17), we have

$$\begin{aligned}&\int_\Omega\int_\Omega \kappa_t(x,y,t)h(y)k(x)\mathrm{d}y\mathrm{d}x + \int_\Omega\int_\Omega \mu\nabla^2 k(x)\kappa(x,y,t)h(y)\mathrm{d}y\mathrm{d}x \\ &+ \int_\Omega\int_\Omega \mu\nabla^2 h(x)\kappa(x,y,t)k(y)\mathrm{d}y\mathrm{d}x = \int_\Omega\int_\Omega k(x)\kappa(x,y,t)\mu^2H^2\sum_{i=1}^N a_i\delta(y-p_i) \\ &\{\sum_{j-1}^N \bar{r}_{ij}a_j\int_\Omega \kappa(p_j,y_1,t)h(y_1)\mathrm{d}y_1\}\mathrm{d}y\mathrm{d}x \\ &- \int_\Omega h(x)k(x)\mathrm{d}x + \int_\Omega \frac{1}{V}h(x)\mathrm{d}x\int_\Omega k(x)\mathrm{d}x\end{aligned}$$

where $R^{-1} = (\bar{r}_{ij})_{1\leq i,j\leq N}$. A simple application of Green's theorem then shows that we can choose κ to satisfy

$$\kappa_t + \mu\nabla_x^2\kappa + \mu\nabla_y^2\kappa + \delta(x - y) - \frac{1}{V} = \sum_{i=1}^{N}\sum_{j=1}^{N} \mu^2 H^2 \kappa(x,p_i,t)\kappa(p_j,y,t)a_i a_j \bar{r}_{ij}$$

subject to the boundary conditions

$$H\kappa(x,y,t) + \lambda \frac{\partial}{\partial n_x}\kappa(x,y,t) = 0, \quad x \in \partial\Omega, \quad y \in \Omega, \quad t \in [0,t_f]$$

$$H\kappa(x,y,t) + \lambda \frac{\partial}{\partial n_y}\kappa(x,y,t) = 0, \quad x \in \Omega, \quad y \in \partial\Omega, \quad t \in [0,t_f]$$

and the final condition

$$\kappa(x,y,t_1) = \delta(x - y) - \frac{1}{V}$$

The optimal control is then

$$u_{\text{op}}(t) = -R^{-1}(a_i \int_\Omega \kappa(p_i,y,t)T(y,t)\text{d}y)_{1\leq i\leq N}\mu H$$

Example 6.3.4

Consider the model of collisionless plasma dynamics described by the Vlasov equation discussed in example 3.5.5. In the latter example we showed that the equations (for both electron and ion species) may be written

$$\dot{f} = Af, f \in L^2(\Omega \times V_c) \oplus L^2(\Omega \times V_c) \triangleq H$$

where A generates a semigroup $T(t)$. If the control is the voltage u placed on a Langmuir probe, then we can write the controlled equation in the form

$$\dot{f} = Af + Bu$$

where

$$Bu = \left(\frac{q_{\text{e}}}{m_{\text{e}}} \nabla\psi \cdot \frac{\partial f_{0\text{e}}}{\partial v}, \frac{q_{\text{i}}}{m_{\text{i}}} \nabla\psi \cdot \frac{\partial f_{0\text{i}}}{\partial v}\right)u$$

for some function ψ. For the cost functional, we choose

$$J(u) = \langle f(t_f) - f_0, f(t_f) - f_0\rangle + \int_0^{t_f} [\langle f(t) - f_0, f(t) - f_0\rangle + u^2]\text{d}t$$

where f_0 (= $(f_{0\text{e}}, f_{0\text{i}})$) is the equilibrium distribution and the inner products are with respect to H.

The feedback control is now given by (6.18) where Q is given by (6.17). An approximation method in terms of eigenspaces is discussed in Banks and

Mousavi-Khalkhali (1982) for the solution of these equations. This example shows that in many physical situations a linear-quadratic tracking problem occurs quite naturally.

Example 6.3.5
As a final example of the linear-quadratic problem, we shall consider the linear multipass process (*cf*. example 3.5.6):

$$\frac{\mathrm{d}x_k(t)}{\mathrm{d}t} = A_0x_k(t) + A_1x_{k-1}(t) + \ldots + A_lx_{k-l}(t) + Bu_k(t)$$

for $k \geq 0$, $t \in [0,\tau]$ where $B \in \mathscr{L}(U,H)$. If we restrict attention to a finite number of passes, $0 \leq k \leq m$, then

$$\dot{x} = \mathscr{A}_m x + \mathscr{B}_m u + f_m \tag{6.23}$$

where $\mathscr{A}_m$ is defined in example 3.5.6,

$$\mathscr{B}_m = \mathrm{diag}(B,\ldots,B) \in \mathscr{B}(\bigoplus_{i=0}^{m} H)$$

and

$$f_m = (A_1x_{-1} + \ldots + A_lx_{-l}, A_2x_{-1} + \ldots + A_lx_{1-l},\ldots,A_lx_{-1},0,0,\ldots)$$

The cost functional for this problem will be

$$J(u) = \langle x(\tau),x(\tau)\rangle_{\mathscr{X}_m} + \int_0^\tau \{\langle x(s),x(s)\rangle_{\mathscr{X}_m} + \langle u(s),u(s)\rangle_{\mathscr{U}_m}\}\mathrm{d}s$$

where we have taken G,D,R to be identity operators for simplicity. The equation (6.23) is not homogeneous and so we put

$$z(t) = x(t) - g(t)$$

where

$$g(t) = \int_0^t \mathscr{T}_m(t-s)f_m(s)\mathrm{d}s$$

and $\mathscr{T}_m$ is the semigroup defined in example 3.5.6. Then we have

$$z(t) - \mathscr{T}_m(t)z(0) + \int_0^t \mathscr{T}_m(t-s)\mathscr{B}_m u(s)\mathrm{d}s \quad (z(0) = x(0))$$

and the cost function becomes

$$J(u) = \|z(\tau) + g(\tau)\|^2_{\mathscr{X}_m} + \int_0^\tau \{\|x(s) + g(s)\|^2_{\mathscr{X}_m} + \|u(s)\|^2_{\mathscr{U}_m}\}\mathrm{d}s$$

This tracking problem has the optimal control

$$u_{\mathrm{op}}(t) = -\mathscr{B}^*_m Q_m(t)z(t) - \mathscr{B}^*_m s(t)$$

where

$$\frac{\mathrm{d}}{\mathrm{d}t}\langle Q_m(t)h_m,k_m\rangle + \langle Q_m(t)h_m,\mathcal{A}_m k_m\rangle + \langle \mathcal{A}_m h_m, Q_m(t)k_m\rangle + \langle h_m,k_m\rangle$$
$$= \langle Q_m(t)\mathcal{B}_m\mathcal{B}^*_m Q_m(t)h_m,k_m\rangle \qquad t\in[0,\tau] \qquad 6.24)$$
$$Q_m(\tau) = I, h_m,k_m \in \mathcal{D}(\mathcal{A}_m)$$

and

$$\frac{\mathrm{d}}{\mathrm{d}t}\langle s(t),h_m\rangle = -\langle s(t),(\mathcal{A}_m - \mathcal{B}_m\mathcal{B}^*_m Q_m(t))h_m\rangle - \langle g(t),h_m\rangle$$
$$s(\tau) = -g(\tau)$$

If

$$h,k \in \bigcap_{i=0}^{l} \mathcal{D}(A_i)$$

then (6.24) becomes

$$\frac{\mathrm{d}}{\mathrm{d}t}\langle Q_{ij}h,k\rangle + \sum_{\alpha=0}^{l}\langle A^*_\alpha Q_{i+\alpha,j}h,k\rangle + \sum_{\alpha=0}^{l}\langle h,A^*_\alpha Q_{j+\alpha,i}k\rangle$$
$$+ \langle h,k\rangle\delta_{ij} = \sum_{\alpha=0}^{m}\langle Q_{i\alpha}BB^*Q_{\alpha j}h,k\rangle \qquad (6.26)$$
$$Q_{ij}(\tau) = I_H\delta_{ij}$$

where we have set

$$h_m = (0,0,\ldots,\underset{j}{h},0,\ldots,0), k_m = (0,0,\ldots,\underset{i}{k},0,\ldots,0)$$

and

$$Q_m = (Q_{ij})_{0\leq i\leq m, 0\leq j\leq m},\ Q_{ij} \in \mathcal{B}(H)$$

(Note that Q_{ij} in (6.26) is interpreted as 0 if $i > m$.) The coupled equations (6.26) will usually have to be integrated numerically as will (6.25). If this done, then the optimal control becomes

$$u_{\mathrm{op}}(t) = -\mathcal{B}^*_m Q_m(t)x(t) - \mathcal{B}^*_m Q(t)g(t) - \mathcal{B}^*_m s(t)$$

This control is noncausal however, since the feedback on the ith pass depends on the states $x_{i+1},\ldots,x_m$ of future passes. This can be obviated by substituting this control in the original equation and solving for $x_i(t)$ in terms of earlier pass states. In fact, we have

$$x_k(t) = T(t)x_k(0) + \sum_{i=1}^{l}\int_0^t T(t-s)A_i x_{k-i}(s)\mathrm{d}s + \int_0^t T(t-s)B(u_{\mathrm{op}})_k(s)\mathrm{d}s$$
$$= T(t)x_k(0) + \sum_{i=1}^{l}\int_0^t T(t-s)A_i x_{k-i}(s)\mathrm{d}s$$

$$-\sum_{j=0}^{m}\int_0^t T(t-s)BB^*Q_{kj}(s)x_j(s)\mathrm{d}s$$
$$-\sum_{j=0}^{m}\int_0^t T(t-s)BB^*Q_{kj}(s)g(s)\mathrm{d}s-\sum_{j=0}^{m}\int_0^t T(t-s)BB^*s(s)\mathrm{d}s$$

We then define the operators $\eta_i,\zeta_{kj}\in\mathcal{B}(L^2([0,\tau];H))$ and $\xi_k^m\in L^2([0,\tau];H)$ by

$$(\eta_i y)(t)=\int_0^t T(t-s)A_i y(s)\mathrm{d}s,\quad 1\leqslant i\leqslant l$$

$$(\zeta_{kj}y)(t)=\int_0^t T(t-s)BB^*Q_{kj}(s)y(s)\mathrm{d}s,\quad 0\leqslant k,j\leqslant m$$

$$\xi_k^m(t)=T(t)x_k(0)-\sum_{j=0}^{m}\int_0^t T(t-s)BB^*Q_{kj}(s)g(s)\mathrm{d}s$$
$$-\sum_{j=0}^{m}\int_0^t T(t-s)BB^*s(s)\mathrm{d}s$$

Then the above equation for $x_k(t)$ can be written

$$Zx=\Xi$$

where Z is the matrix operator defined by

$$\begin{pmatrix}(I+\zeta_{00}) & \zeta_{01} & \cdots & \zeta_{0m}\\ \zeta_{10} & (I+\zeta_{11}) & \cdots & \zeta_{1m}\\ \cdots\cdots & \cdots\cdots & \cdots\cdots & \cdots\cdots\\ \zeta_{m0} & \cdots\cdots & \cdots & (I+\zeta_{mm})\end{pmatrix}+\begin{pmatrix}0 & & & & & \\ -\eta_1 & \ddots & & & \mathbf{0} & \\ -\eta_2 & \ddots & \ddots & & & \\ \vdots & \ddots & \ddots & \ddots & & \\ -\eta_l & & \ddots & \ddots & \ddots & \\ 0 & -\eta_l & \cdots & -\eta_2 & -\eta_1 & 0\end{pmatrix}$$

(for $m>l$) and $\Xi=(\xi_0^m,\xi_1^m,\ldots,\xi_m^m)^T$. Z can now be reduced to a lower triangular matrix operator in the normal way and we obtain a causal controller. (For more details, see Banks, 1982, where the specific example

$$\frac{\partial z_k}{\partial t}(t,x)=\frac{\partial^2 z_k}{\partial x^2}(t,x)+z_{k-1}(t,x)+u_k(t,x)$$

is completely solved.)

6.4 The receding horizon principle

In this section we shall generalise the finite dimensional receding horizon principle mentioned in the introduction to infinite dimensional systems. The discussion will be somewhat informal because of the technical difficulties involved and the shortage of space available here. A rigorous derivation is given by Banks (1983). In Section 6.2 we showed that the optimal control is determined by solving the Riccati equation (6.17); i.e.

$$\dot{Q} = -A^*Q - QA + QBR^{-1}B^*Q \tag{6.27}$$

where this equation must be interpreted in the 'weak' inner product form of (6.17). Then, we have, using the system dynamics (6.7),

$$\dot{x} = Ax - BR^{-1}B^*Qx$$

and so

$$\begin{aligned} Q\dot{x} &= QAx - QBR^{-1}B^*Qx \\ &= -\dot{Q}x - A^*Qx \end{aligned}$$

Hence,

$$\overline{\dot{Q}x} = -A^*Qx$$

and so

$$Qx(t) = T^*(-(t - t_0))Q(t_0)x_0$$

It follows that the optimal control may be written in the open-loop form

$$u_{\text{op}} = -R^{-1}B^*T^*(t_0 - t)Q(t_0)x_0 \tag{6.28}$$

where we assume, temporarily, that T is a group (i.e. is defined for t) and T^* is the dual group of T.

To derive the receding horizon control, consider the equation

$$\dot{W}(t) = AW + WA^* - BR^{-1}B^*, t \in [0,\tau], W(\tau) = \frac{I}{\alpha} \tag{6.29}$$

which has the unique solution

$$W(t) = T(t-\tau)W(\tau)T^*(t-\tau) + \int_0^\tau T(-s)BR^{-1}B^*T^*(-s)\mathrm{d}s \tag{6.30}$$

It is easy to show that if $\alpha > 0$, then the solutions of (6.27) (with $G = \alpha I$ in (6.9)) and (6.29) are related by $Q(t)W(t) = I$, $W(t)Q(t) = I$, $t \in [0,\tau]$ and so Q is invertible on $[0,\tau]$. Now, using theorem 5.2.3, if the system is exactly controllable, then it can be shown that we may let $\alpha \to \infty$ and we find that $W(t)$ is invertible on $[0,\tau[$. The control which minimises the cost functional

$$J(u) = \int_{t_0}^{t_0+T} \langle u,Ru\rangle \mathrm{d}t$$

subject to

$$x(t_0 + T) = 0$$

is therefore

$$u_{\mathrm{op}} = -R^{-1}B^{*}T^{*}(t_0 - t)W^{-1}(0)x_0 \tag{6.31}$$

(Note that, in the above, it has been convenient to reverse time in equation (6.30) compared to (6.5*a*), which accounts for the apparent difference in the control.) The control (6.31) is, of course, open loop and so will not be stable to perturbations. In the receding horizon philosophy we argue that this control should be applied as if, at each instant of time, we were just beginning a new minimisation over the next τ seconds. Hence, we replace (6.31) by the closed-loop control

$$u^{*} = -R^{-1}B^{*}W^{-1}(0)x(t) \tag{6.32}$$

We can then make this (suboptimal) control nonlinear by allowing the horizon time τ to depend on the state x ($W(0)$ depends on τ and so u^{*} will be nonlinear in x via the function W^{-1}). With a slightly more delicate argument we can show that (6.32) is also valid in the semigroup case (see Banks, 1983).

Example 6.4.1
Consider the controlled heat equation

$$x_t = x_{\xi\xi} + u(t,\xi)$$
$$x_{\xi}(0,t) = 0 = x_{\xi}(1,t)$$

as in example 6.3.1. Writing as before

$$Q(t)h = \sum_{i=0}^{\infty}\sum_{j=0}^{\infty} q_{ij}(t)\phi_j\langle h,\mu_i\rangle h,\phi_i\rangle$$

we find (using 6.17) that

$$q_{00}(t) = (t_f - t + 1/\alpha)^{-1}$$

$$q_{ii}(t) = \left\{\left(\frac{1}{\alpha} + \frac{1}{2\pi^2 i^2}\right)e^{2\pi^2 i^2(t_f - t)} - \frac{1}{2\pi^2 i^2}\right\}^{-1}, \; i \geqslant 1$$

$$q_{ij}(t) = 0, \; i \neq j$$

where we have taken $G = \alpha I$, $D = 0$.
Hence,

$$w_{ij} = (Q^{-1})_{ij} = \delta_{ij}\left\{\left(\frac{1}{\alpha} + \frac{1}{2\pi^2 i^2}\right) e^{\,2\pi^2 i(t_f - t)} - \frac{1}{2\pi^2 i^2}\right\}, \; i \neq 0$$

$$w_{00} = t_f - t - \frac{1}{\alpha}$$

and so, as expected, we can take the limit $\alpha \to \infty$ and we have

$$w_{ii} = \frac{1}{2\pi^2 i^2}(e^{2\pi^2 i^2(t_f - t)} - 1)$$

$$w_{00} = t_f - t$$

It follows that the controlled system has equations

$$\dot{x}_0(t) = -\frac{1}{t_f} x_0(t)$$

$$\dot{x}_i(t) = -(\pi i)^2 x_i(t) - \frac{2\pi^2 i^2}{\{\exp(2\pi^2 i^2 t_f) - 1\}} x_i(t)$$

We can then choose t_f to depend on x; for example, we may take $t_f = \|x\|_{L^2}^{-2}$. This system should be compared with the linear-quadratic system in example 6.3.1.

Example 6.4.2
Returning now to the wave equation in example 6.3.2, we consider the cost functional

$$J(u) = \int_0^{t_f} \int_0^1 u^2(\xi,t)\mathrm{d}\xi \mathrm{d}t$$

Then using the above theory it is quite easy to show that the receding horizon control has the ith component

$$\langle u^*,\phi_i\rangle_{L^2} = 2\left\{(i\pi)^2\left(t_f^2 - \frac{\sin^2 2i\pi t_f}{4(i\pi)^2}\right) - \tfrac{1}{4}(\cos 2i\pi t_f - 1)^2\right\}^{-1}$$

$$\times \left\{\tfrac{1}{2}(\cos 2i\pi t_f - 1)\,\langle x_\xi(t),\phi_i\rangle_{L^2} - (i\pi)^2\left(t_f - \frac{\sin 2i\pi t_f}{2i\pi}\right)\langle x_t,\phi_i\rangle_{L^2}\right\}$$

with respect to the basis $\phi_i(\xi) = \sqrt{2}\sin i\pi\xi$.

Note that if it_f is an integer, we get

$$\langle u^*,\phi_i\rangle_{L^2} = -\frac{2}{t_f}\langle x_t,\phi_i\rangle_{L^2}$$

and if $t_f = 4$, we obtain the linear-quadratic solution. The receding horizon control is therefore much more flexible than the LQR version. Finally, we can make the control nonlinear by choosing, for example,

$$t_f = \|(x,t_t)\|_{\mathscr{X}}^{-1}$$

6.5 Time optimal control

In this section we shall consider the time optimal control problem for the wave equation (as discussed by Fattorini, 1977). This system will illustrate the effect of 'hard' control constraints and the generalisation of the maximum principle for distributed systems. The derivation involves, basically, an application of the geometric form of the Hahn-Banach theorem (*cf.* theorem 2.2.12) which states that a nonvoid convex set C in a Hilbert space H can be separated from any point h not in the set, i.e. there exists a hyperplane in H such that C and h lie on opposite sides of this hyperplane. Consider then the wave equation

$$\frac{\partial u}{\partial t} = c^2 \sum_{k=1}^{p} \frac{\partial^2 u}{\partial x_k^2} + f = c^2 \Delta u + f, x \in \Omega \tag{6.33}$$

with $u(x) = 0$ for $x \in \Gamma = \partial\Omega$, and f is the control function. A simple application of theorem 5.2.3 shows that this system is exactly controllable. We assume that f is bounded, i.e.

$$\|f(t)\|_{L^2(\Omega)} \leqslant C \tag{6.34}$$

If $A = c^2\Delta$, then a simple generalisation of example 3.5.3 shows that $\begin{pmatrix} 0 & I \\ A & 0 \end{pmatrix}$ generates a semigroup $T(t)$ on $\mathcal{H} = \mathcal{D}((-A^{1/2}) \oplus L^2(\Omega)$, and we can write

$$T(t) = \begin{pmatrix} C(t) & S(t) \\ AS(t) & C(t) \end{pmatrix}$$

where C and S are the 'cosine' and 'sine' functions defined by

$$C(t) = c(t,A),\ S(t) = s(t,A)$$

where

$$c(t,\lambda) = \cos[(-\lambda)^{1/2}t],\ s(t,\lambda) = (-\lambda)^{1/2} \sin[(-\lambda)^{1/2}t]$$

Then the equation (6.33) (written as a pair of first-order equations) has the 'mild' solution

$$w(t) = \begin{pmatrix} u(t) \\ u'(t) \end{pmatrix} = \begin{pmatrix} C(t)u_0 + S(t)u_1 + \int_0^t S(t-s)f(s)\mathrm{d}s \\ AS(t)u_0 + C(t)u_1 + \int_0^t C(t-s)f(s)\mathrm{d}s \end{pmatrix}$$

where $u(0) = u_0$, $u'(0) = u_1$.

We must now show that a control which drives the system between two given states in minimum time exists. A function f satisfying (6.34) which is strongly measurable will be called an *admissible control*.

Theorem 6.5.1
If $\boldsymbol{u} = (u_0,u_1)$, $\boldsymbol{v} = (v_0,v_1) \in H$, then there exists an admissible control driving $\boldsymbol{u}$ to $\boldsymbol{v}$ in minimum time.

Proof: Let

$$S = \{t \in \mathcal{R}_+ : \exists \text{ an admissible control driving } \boldsymbol{u} \text{ to } \boldsymbol{v} \text{ in time } t\}$$

and define

$$t_0 = \inf S$$

Then there exists a sequence $\{f_n\}$ of admissible controls driving $\boldsymbol{u}$ to $\boldsymbol{v}$ in times $t_1 \geqslant t_2 \geqslant \ldots \geqslant t_0$ such that $t_n \to t_0$. Extending by zero, the controls f_n belong to $L^2(0,t_1;L^2(\Omega))$. By theorem 2.5.7, there is a subsequence (again denoted by $\{f_n\}$) such that $f_n \to f_0$ (weakly) and $f_0(s) = 0$ for $t \geqslant t_0$. Now we have

$$\int_0^{t_n} T(t_n - t)f_n(t)\mathrm{d}t = -\begin{pmatrix} C(t_n)u_0 + S(t_n)u_1 - v_0 \\ AS(t_n)u_0 + C(t_n)u_1 - v_1 \end{pmatrix}$$

and so by the weak convergence of f_n it follows that f_0 drives $\boldsymbol{u}$ to $\boldsymbol{v}$. □

Knowing that minimum time control exists we now define the set of points in $\mathcal{H}$ which can be reached from $\boldsymbol{u} = 0$ with admissible controls in time t_0; i.e. the set

$$\Delta = \{\mathcal{H} \ni (u_1,u_2) = \int_0^{t_0} T(t_0 - s)f(s)\mathrm{d}s \colon f \text{ admissible}\}$$

Then, it can be shown (Fattorini, 1964; 1977):

(*a*) $(\Delta)^0$ is nonempty;
(*b*) $(w_0,w_1) = (v_0,v_1) - (C(t_0)u_0 + S(t_0)u_1, AS(t_0)u_0 + C(t_0)u_1)$ is a boundary point of Δ.

Now, by the Hahn-Banach theorem, there is a nonzero continuous linear functional l on $\mathcal{H}$ such that

$$l((u,u')) < l((w_0,w_1)) \tag{6.35}$$

for all $(u,u') \in \Delta$. Now l must be of the form

$$l((u,u')) = \langle h,(-A)^{1/2}u\rangle + \langle h',u'\rangle$$

for some $h,h' \in L^2(\Omega)$ and so (6.35) becomes

$$\begin{aligned} &\int_0^{t_0} \langle(-A)^{1/2}S(t_0 - t)h,f(t)\rangle\mathrm{d}t + \int_0^{t_0} \langle C(t_0 - t)h',f(t)\rangle\mathrm{d}t \\ \leqslant &\int_0^{t_0} \langle(-A)^{1/2}S(t_0 - t)h,f_0(t)\rangle\mathrm{d}t + \int_0^{t_0} \langle C(t_0 - t)h',f_0(t)\rangle\mathrm{d}t \end{aligned}$$

and so

$$\begin{aligned} &\langle(-A)^{1/2}4S(t_0 - t)h + C(t_0 - t)h',f_0(t)\rangle \\ &= \sup_{\|f\|\leqslant 1} \langle(-A)^{1/2}S(t_0 - t)h + C(t_0 - t)h',f\rangle \end{aligned} \tag{6.36}$$

a.e. in $0 \leqslant t \leqslant t_0$, for some h,h' (not both zero) and for any control f_0 driving

(u_0,u_1) to (v_0,v_1) in minimum time t_0 (we have taken the constant $C = 1$ for simplicity). When

$$\Phi(h,h';t) \triangleq (-A)^{1/2}S(t_0 - t)h + C(t_0 - t)h' = 0 \tag{6.37}$$

the equation (6.36) gives no information about $f_0(t)$. However, it can be shown that there can only be a finite number of points $t \in [0,t_0]$ at which (6.37) holds. Hence, we have the necessary condition

$$f_0(t) = \frac{\Phi(h,h';t)}{\|\Phi(h,h';t)\|}, \quad 0 \leqslant t \leqslant t_0$$

except for a finite number of values of t.

The time optimal problem for parabolic systems has been studied by Washburn (1979). For some other aspects of nonlinear control see Barbu (1981).

Chapter 7

System theory approach and the infinite dimensional root locus

7.1 Introduction

In the previous chapters of this book we have been concerned mainly with the infinite dimensional generalisations of some of the control theoretical results of finite dimensional systems in the state space formulation. The emphasis in this last chapter will be on the systems theory (or transfer function) approach to distributed systems and in particular on the infinite dimensional root locus; it will necessarily be less complete than the state-space theory, since, as with finite dimensional multivariable systems, the frequency domain techniques have not, until recently, attracted as much attention as the state-space methods (for an account of finite dimensional multivariable theory see, for example, Rosenbrock, 1974; Postlethwaite and MacFarlane 1979; Owens, 1978).

There are, of course, many instances in the literature where the classical techniques have been applied to distributed systems by joint Laplace and Fourier transforms resulting in nonrational transfer functions. However, there has been little attempt at justifying the results rigorously. Indeed, in this chapter we shall show that even the root locus of an infinite dimensional system with a bounded realisation can exhibit some surprising phenomena, unless they are interpreted in the correct way, when they appear as natural generalisations of the finite dimensional root locus. When the system has a discrete spectrum as is the case, for example, with the heat conduction equation, then the root locus has a branch on each spectral point and the behaviour of the root locus is then much easier to relate to the classical case. It is the general case which is more interesting as we shall see, and the study of the distributed root locus will just appear as an application of spectral theory.

7.2 Transfer function theory

In this section we shall derive the generalisation of the rational transfer function for finite dimensional systems to distributed systems. Consider, then, a system

$$\begin{aligned} \dot{x} &= Ax + Bu \\ y &= Cx \end{aligned} \tag{7.1}$$

defined on a Hilbert space H. It will be sufficient for our purposes to suppose that $A:\mathscr{D}(A) \subseteq H \to H$ generates a strongly continuous semigroup, $B \in \mathscr{B}(U,H)$ and that C is an A-bounded operator (*cf.* (3.13)) with

$$C:H \supseteq \mathscr{D}(C) \supseteq \mathscr{D}(A) \to 0$$

for some input and output spaces U,0, respectively.

The transfer function is traditionally defined by assuming zero initial condition $x(0) = 0$ in (7.1). Then, by the variation of constants formula,

$$x(t) = \int_0^t T(t - \tau)Bu(\tau)d\tau$$

Hence, using proposition 3.2.5, we obtain, by Laplace transformation,

$$X(s) = R(s;A)BU(s), \operatorname{Re} s > \omega(= \inf t^{-1} \log \|T(t)\|) \tag{7.2}$$

where

$$X = \mathscr{L}(x), \; U = \mathscr{L}(u)$$

However, by (7.1) we have $y = Cx$. Now, by (7.2), $X(s) \in \mathscr{D}(A)$ for all s with $\operatorname{Re} s > \omega$ and therefore, since C is A-bounded it follows that $Y(s) = CX(s)$, $\operatorname{Re} s > \omega$. Hence,

$$Y(s) = CR(s;A)BU(s) \tag{7.3}$$

for all $s \in \rho(A)$, by analytic continuation. We therefore introduce the following definition.

Definition 7.2.1
The *(open-loop) transfer function* of the system (7.1) is defined by

$$G(s) = CR(s;A)B, s \in \rho(A) \tag{7.4}$$

In the case of finite dimensional systems the open-loop transfer function $G(s)$ is, of course, a matrix of rational functions; in contrast to this, the transfer function of a distributed system can be irrational and may be nonanalytic at isolated points, along cuts joining branch points or even on Jordan regions of the plane (i.e. regions surrounded by a Jordan curve). The latter may happen when the input and output spaces are infinite dimensional, as we shall see.

Example 7.2.2
Consider the problem of heat flow in a bar of unit length with control at a point x_1 and measurement at a point x_2. Then we have the system described

by the equations

$$\frac{\partial\phi}{\partial t} = \frac{\partial^2\phi}{\partial x} + \delta(x - x_1)u, \quad \phi(0,t) = \phi(1,t) = 0$$

$$y = \langle\delta_{x_2},\phi\rangle$$

(This corresponds to heat injection of magnitude u at x_1, and is given by the condition $-[\phi_x]_{x_1} = u$, i.e. change of the derivative of ϕ at x_1 equals u; see Curtain and Pritchard, 1978). We can then choose the state space to be $H^{-1/2-\varepsilon}(0,1)$. Now,

$$(sI - A)^{-1}\phi = \int_0^\infty \exp(-st)T(t)\phi dt$$

where $T(t)$ is the semigroup given in example 3.5.1 and A is the operator $\partial^2/\partial x^2$ with obvious domain. Hence,

$$(sI - A)^{-1}\phi = \sum_{n=1}^{\infty} 2\int_0^\infty e^{-(s+n^2\pi^2)t}dt \,.\, \sin(n\pi x) \int_0^1 \sin(n\pi\rho)\phi(\rho)d\rho$$

$$= \sum_{n=1}^{\infty} \frac{2}{s + n^2\pi^2} \sin(n\pi x) \int_0^1 \sin(n\pi\rho)\phi(\rho)d\rho$$

and so

$$(sI - A)^{-1}Bu = \sum_{n=1}^{\infty} \frac{2}{s + n^2\pi^2} \sin(n\pi x)\sin(n\pi x_1)u$$

It follows that the open-loop transfer function is

$$G_0(s) = C(sI - A)^{-1}B = \sum_{n=1}^{\infty} \frac{2}{s + n^2\pi^2} \sin(n\pi x_2)\sin(n\pi x_1) \tag{7.5}$$

Clearly, $G_0(s)$ is analytic in $\mathscr{C}$ except at a countable number of discrete first-order poles $s = -n^2\pi^2$, as one would expect from the spectral theory of operators with compact resolvent (Chapter 2) and the spectral minimality of such operators (Section 5.4), provided the system is controllable and observable; i.e. x_1,x_2 irrational.

It is interesting to note that we may write

$$G_0(s) = \frac{\sum_{m=1}^{\infty} \prod_{n\neq m=1}^{\infty} (2/m^2\pi^2)(s/n^2\pi^2 + 1)\sin(m\pi x_2)\sin(m\pi x_1)}{\prod_{n=1}^{\infty}(s/n^2\pi^2 + 1)}$$

However, the denominator is a well known infinite product (Whittaker and

Watson, 1965) given by

$$\prod_{n=1}^{\infty} (s/n^2\pi^2 + 1) = \frac{1}{\sqrt{s}} \sinh \sqrt{s}$$

and so

$$G_0(s) = \frac{\sqrt{s}n(s)}{\sinh \sqrt{s}} \tag{7.6}$$

where

$$n(s) = \sum_{m=1}^{\infty} \frac{2}{m^2\pi^2} \sin (m\pi x_2) \sin (m\pi x_1) \prod_{n \neq m=1}^{\infty} (s/n^2\pi^2 + 1)$$

Note that $n(s)$ has no zero at a pole of $G_0(s)$, so that we confirm the spectral minimality.

One can obtain a closed form for $G_0(s)$ by taking Laplace transforms of the equation

$$\frac{\partial \phi}{\partial t} = \frac{\partial^2 \phi}{\partial x^2}, \ \phi(0,t) = \phi(1,t) = 0$$

with the condition $-[\phi_x]_{x_1} = u$. Then if $0 < x_2 < x_1$, we obtain

$$G_0(s) = \frac{-P_1(x_1)\sinh \sqrt{s}x_2}{\sqrt{s}(S(x_1)P_2(x_1) - C(x_1)P_2(x_1))} \tag{7.7}$$

where

$$S(x_1) = \sinh \sqrt{s}x_1$$

$$C(x_1) = \cosh \sqrt{s}x_1$$

$$P_1(x_1) = \frac{\sinh \sqrt{s}x_1 - \sinh \sqrt{s} \cosh \sqrt{s}x_1}{\cosh \sqrt{s}}$$

$$P_2(x_1) = \sinh \sqrt{s}x_1 - \frac{\sinh \sqrt{s} \sinh \sqrt{s}x_1}{\cosh \sqrt{s}}$$

Example 7.2.3

In just the same way as in the last example, it is easy to see that the transfer function of the damped wave equation

$$\frac{\partial^2 \phi}{\partial t^2} + 2\xi\omega_0 \frac{\partial \phi}{\partial t} + \omega_0^2\phi = c^2 \frac{\partial^2 \phi}{\partial x^2} + \delta(x - x_1)u(t)$$

$$y = \langle \delta_{x_2}, \phi \rangle, \ \phi(0,t) = \phi(1,t) = 0$$

(with boundary control and observation at x_1,x_2, respectively) is

$$G_0(s) = 2\sum_{k=1}^{\infty} \frac{(\sin k\pi x_1)(\sin k\pi x_2)}{s^2 + 2\zeta\omega_0 s + \omega_0^2 + \omega_k^2} \tag{7.8}$$

where

$$\omega_k = k\pi c,\ k = 1,2,\ldots$$

If $\omega_0 = \xi = 0$, then we have

$$G_0(s) = \frac{s/c}{\sinh\,(s/c)}\, n(s)$$

where

$$n(s) = 2\sum_{m=1}^{\infty} \frac{1}{m^2\pi^2c^2}\,(\sin m\pi x_1)\,(\sin m\pi x_2) \prod_{n\neq m=1}^{\infty} (s^2/n^2\pi^2c^2 + 1)$$

7.3 Root locus of distributed systems

In this section we shall discuss the root locus of infinite dimensional systems. We shall consider first systems with a bounded realisation since a fairly complete theory in such cases has recently been developed (Banks and Abbasi-Ghelmansarai, 1983*b*). To begin, we shall relate the classical realisation (5.2) to the realisation of a bounded system in terms of the right-shift operator, as in (5.26). Suppose that a finite dimensional system is defined by the rational transfer function

$$G(s) = \frac{\sum_{i=0}^{m} a_i s^i}{\left(\sum_{i=0}^{n-1} b_i s^i + s^n\right)},\ m < n \tag{7.9}$$

Then, as is well known, G has the canonical realisation

$$A = \begin{bmatrix} 0 & 1 & & & \\ & 0 & 1 & & \\ & & \cdots\cdots & & \\ & & & 0 & 1 \\ b_0 & b_1 & \ldots & b_{n-2} & b_{n-1} \end{bmatrix}$$

$$b = (0,0,\ldots,0,1)$$

$$c = (a_0,a_1,\ldots,a_m,0,\ldots,0)$$

We can then ask how this relates to the realisation in terms of the right-shift operator given in (5.26). In fact,

$$G(s) = \sum_{i=0}^{\infty} c_i s^{(m-n)-i}$$

where we define inductively,

$$c_0 = a_m$$

$$c_k = a_{m-k} - \left(\sum_{\substack{0 \leqslant i < n \\ n-k-i \leqslant 0}} b_i c_{-n+k+i} \right), \ k \geqslant 1$$

Hence we can realise (7.9) also by

$$\begin{aligned} A &= kU_r \\ b &= (1,0,0,\ldots) \\ c &= (\underbrace{0,0,\ldots,0}_{n-m-1},c_0/k^{n-m-1},c_1/k^{n-m},\ldots) \end{aligned} \tag{7.10}$$

in l^2, where U_r is the right shift operator and

$$k > \max_{s \in \sigma(G)} |s|$$

Note that in representing a finite dimensional system in terms of the right-shift operator, we have obtained a non-minimal realisation. The following theorem of classical complex function theory is useful in connection with the limits of sequences of finite dimensional realisations (see, for example, Rudin, 1974).

Theorem 7.3.1 (Runge's Theorem)
Let Ω be an open set in $\mathscr{C}$, let A be a set which has one point in each component of $S^2 - \Omega$ and assume that f is analytic in Ω. Then there exists a sequence $\{R_n(s)\}$ of rational functions, with poles in A, such that $R_n \to f$ uniformly on compact sets in Ω. □

Here $S^2 \triangleq \mathscr{C} \cup \{\infty\}$ is the Riemann sphere, i.e. the compactified complex plane. In order to apply this theorem, we rewrite G in the form

$$G(s) = \frac{\prod_{i=1}^{m} (s - z_i)}{\prod_{i=1}^{n} (s - p_i)} \tag{7.11}$$

Then,

$$a_m = b_n = 1$$

and

$$a_i = (-1)^{m-i} \sigma_{m-i}(z_1,\ldots,z_m)$$

$$b_i = (-1)^{n-i} \sigma_{n-i}(p_1,\ldots,p_n)$$

where σ_i is the ith elementary symmetric function. Hence, the realisation of (7.11) is given by (7.10) with

$$
\begin{aligned}
c_0 &= 1 \\
c_k &= (-1)^{m-k}\sigma_{m-k}(z_1,\ldots,z_m) \\
&\quad - \sum_{\substack{0\leq i<n \\ n-k-i\leq 0}} \{(-1)^{n-i}\sigma_{n-i}(p_1,\ldots,p_n)c_{-n+k+i}\}
\end{aligned}
\tag{7.12}
$$

and

$$k > \max_{1\leq i\leq n} \{|p_i|\}$$

Consider a single-input–single-output system with transfer function $G(s)$ and suppose that it has a bounded realisation. Then $\sigma(G)$ contains poles and branch cuts, the latter necessarily being in the finite plane since G is analytic at $s = \infty$. $\sigma(G)$ is closed, so we may apply Runge's theorem with $A = \sigma(G) \cup \{\infty\}$. However, $|G(s)| \to 0$ as $|s| \to \infty$, and so we may clearly assume that the rational approximations just have poles in $\sigma(G)$. Hence, we take $A = \sigma(G)$ in theorem 7.3.1 and we obtain a sequence of rational functions R_n such that $R_n \to G(s)$ uniformly on compact sets. Suppose that

$$R_n = \frac{\prod_{i=1}^{m_n} (s - z_i^n)}{\prod_{i=1}^{n} (s - p_i^n)} \tag{7.13}$$

where we have assumed, without loss of generality, that R_n has n poles, $p_i^n (1 \leq i \leq n)$ and m_n $(<n)$ zeros $z_i^n (1 \leq i \leq m_n)$. Hence, we see that $G(s)$ has a realisation in the form (5.26) which can be obtained as a limit of a sequence of realisations of the form (7.10), with c_i^j given by

$$c_0^n = 1$$

$$c_k^n = (-1)^{m_n-k}\sigma_{m_n-k}(z_1^n,\ldots,z_{m_n}^n) - \sum_{\substack{0\leq i<n \\ n-k-i\leq 0}} \{(-1)^{n-i}\sigma_{n-i}(p_1^n,\ldots,p_n^n)c_{-n+k+i}^n\}$$

for $n = 1,2,\ldots$.

For example, if

$$G(s) = \frac{1}{\sqrt{(s^2+1)}}$$

then

$$\sigma(G) = \{s \in \mathscr{C}: \mathrm{Re}\, s = 0,\ |\mathrm{Im}\, s| \leq 1\} \triangleq l$$

and so G consists of the branch points $\pm i$ together with the cut along the imaginary axis joining these points. Therefore, we can find a sequence $\{G_n\}$

of rational functions with n poles on l such that

$$G_n \to G \text{ uniformly on compacta}$$

We are now in a position to consider the root locus of an infinite dimensional system with bounded realisation. (The general theory of root locus for systems with unbounded realisation remains to be clarified; although for systems with compact resolvent the root locus properties are similar to those developed below.) Consider then the system

$$\begin{aligned} \dot{x} &= Ax + kbu \\ y &= Cx \end{aligned} \tag{7.14}$$

where A,b,c are bounded operators belonging, respectively, to $\mathscr{L}(H,H)$, $\mathscr{L}(\mathscr{R},H)$ and $\mathscr{L}(H,\mathscr{R})$ for some Hilbert space H. Suppose that

$$\sigma(A) = \bigcup_{i=1}^{n} \sigma_i$$

where each σ_i is a spectral set of A. Then we can write

$$H = \bigoplus_{i=1}^{n} H_{\sigma_i}$$

$$A_{\sigma_i} = A|_{H_{\sigma_i}}$$

and

$$bu = (b_1u,\ldots,b_nu),\ cx = c_1x_1 + \ldots + c_nx_n$$

where $x_i \in H_{\sigma_i}$, $b_iu \in H_{\sigma_i}$ and $c_i = c|_{H_{\sigma_i}}$. Hence

$$\begin{aligned} \dot{x}_i &= A_ix_i + kb_iu \\ y &= \Sigma c_ix_i \end{aligned} \tag{7.15}$$

where we have written $A_i \triangleq A_{\sigma_i}$.

Taking Laplace transforms in the usual way, we obtain

$$sX_i(s) = A_iX_i(s) + kb_iU(s)$$

$$Y(s) = \Sigma c_iX_i(s)$$

where

$$X_i(s) = \mathscr{L}(x_i(t)),\ U(s) = \mathscr{L}(u(t)),\ Y(s) = \mathscr{L}(y(t))$$

If $u = v - y$, then

$$(1 + k\sum_{i=1}^{n} c_iR(s;A_i)b_i)Y(s) = k(\sum_{i=1}^{n} c_iR(s;A_i)b_i)V(s) \tag{7.16}$$

provided that $s \notin \sigma(A_i) = \sigma_i$, $1 \leqslant i \leqslant n$. Hence, if

$$s \notin \{\lambda \in \mathscr{C}: 1 + k\sum_{i=1}^{n} c_iR(s;A_i)b_i = 0\}$$

we have

$$Y(s) = \left\{ \frac{k(\sum_{i=1}^{n} c_i R(s;A_i)b_i)}{1 + k \sum_{i=1}^{n} c_i R(s;A_i)b_i} \right\} V(s)$$

$$= G(s)V(s)$$

where

$$G(s) \triangleq \frac{k(\sum_{i=1}^{n} c_i R(s;A_i)b_i)}{1 + k \sum_{i=1}^{n} c_i R(s;A_i)b_i} \tag{7.17}$$

G is called the closed-loop transfer function of (7.14). If we want to emphasise that G depends on the feedback gain k we shall write $G(s;k)$. We can now define the *root locus* L of the system (7.14) as the set

$$L = \bigcup_{k \geq 0} \{s \in \mathscr{C}: G(s;k) \text{ is not analytic at } s\}$$

Note that for finite dimensional systems, the decomposition (7.15) just corresponds to the Jordan decomposition of A, in which case the open-loop transfer function

$$G_0(s) = \sum_{i=1}^{n} c_i R(s;A_i)b_i$$

is expressed as a partial fraction expansion. We shall see shortly that this can be generalised somewhat.

An important concept in finite dimensional systems theory is that of transmission zero, for it is to these points that bounded parts of the root locus are attracted as $k \to \infty$. In the case of infinite dimensional systems with compact resolvent, Pohjolainen (1981) has defined the transmission zeros to be the finite eigenvalues λ of $A + kBC$ as $k \to \infty$. It is then easy to show that they must satisfy the condition

$$\lambda \notin \sigma(A) \text{ and } \det [C(A - \lambda I)^{-1}B] = 0 \tag{7.18}$$

These systems are spectral minimal and so the definition is appropriate here. However, in general there is no such simple relation between the spectrum of $A + kBC$ and the zeros of $\det [C(A - \lambda I)^{-1}B]$. Hence, we revert to (7.18) as a preliminary definition for the transmission zeros. This is not quite enough, however, since (for single-input–single-output systems) we must not only include zeros of $c(A - \lambda I)^{-1}b$ but also any cut in the plane which makes $c(A - \lambda I)^{-1}b$ single valued. Hence, we introduce the following definition.

Definition 7.3.2
The *transmission zero set* ζ for the system (A,b,c) is defined as

$$\zeta(G_0) = \{s \in \mathscr{C} : s \notin \sigma(A),\ G_0^{-1} \text{ is not analytic at } s\} \tag{7.19}$$

where $G_0 = c(A - sI)^{-1}b$ is the open-loop transfer function.

Definition 7.3.3
A connected component of $\zeta(G_0)$ is called a *generalised open-loop zero*, while a connected component of $\sigma(G_0)$ is called a *generalised open-loop pole*. Similarly, a connected component of $\sigma(G)$ is a *generalised closed-loop pole*.

It should be noted that the transmission zero set (and the generalised pole set) is not uniquely specified by definition 7.3.2, since the cuts are to some extent arbitrary. However, we shall understand the definition in the sense that the necessary cuts are chosen and fixed in the discussion. Now we know in finite dimensions that the transfer function can be written in the form

$$G(s) = \frac{\prod_{i=1}^{m}(s - z_i)}{\prod_{i=1}^{n}(s - p_i)}$$

This can be generalised as follows.

Theorem 7.3.4
Suppose that, for a system with open-loop transfer function $G_0(s)$, all the branch points are algebraic. Then we may write G_0 in the form

$$G_0(s) = \left\{\frac{\prod_{i=1}^{m_1}\Psi_i(s)}{\prod_{i=1}^{n_1}\Phi_i(s)}\right\} R(s) \tag{7.20a}$$

where Φ_i and Ψ_i are each of the form $\sqrt[n]{\{s^n + a_1 s^{n-1} + \ldots + a_n\}}$ and $R(s)$ is a rational function corresponding to classical poles and zeros. Moreover, if R has m_2 zeros and n_2 poles, then $m_1 + m_2 < n_1 + n_2$ and

$$\Psi_i(s) - s, \frac{s}{\Phi_i(s)} - 1 \text{ are analytic at } \infty \text{ and vanish there} \tag{7.20b}$$

(For a proof see Banks and Abbasi-Ghelmansarai, 1983*b*). □

Definition 7.3.5
If the system (7.14) is such that each σ_i contains a single generalised pole, then the spectral representation (7.15) of the system will be called *simple*.

A simple consequence of (7.17) and theorem 7.3.4 is the following.

Corollary 7.3.6
Under the conditions of theorem 7.3.4, if the system (7.14) has a simple spectral representation (7.15), then we may write

$$G_0(s) = \sum_{i=1}^{n} h_i(s)/\Phi_i(s) \tag{7.21}$$

where each h_i is holomorphic in $\mathscr{C} \cup \{\infty\}$. □

The expression (7.21) is the *generalised partial fraction expansion.*

Theorem 7.3.7
Suppose that the open-loop transfer function $G_0(s)$ of the system (7.14) is written in the form of (7.20*a*) with $m = m_1 + m_2$ and $n = n_1 + n_2$. Then we have

$$G_0(s) = k\left(\frac{a_{n-m}}{s^{n-m}} + \frac{a_{n-m+1}}{s^{n-m+1}} + \dots\right)$$

for some $a_{n-m} \neq 0$ in a neighbourhood of $s = \infty$.

Proof: This follows from (7.20*b*) since each term Ψ_i, Φ_i has an expansion

$$\sum_{i=1}^{\infty} \alpha_i s^i$$

with $\alpha_1 \neq 0$. □

Corollary 7.3.8
If the system (7.14) has m generalised zeros and n generalised poles, then m of the poles converge, as $k \to \infty$ to the generalised zeros and the remaining $n - m$ generalised poles tend to ∞ with asymptotic directions given by the angles $\theta = l\pi/(n - m)$ and which intersect the x-axis at

$$\sigma_0 = \left(\frac{a_{n-m+1}}{a_{n-m}}\right)/(n - m). \quad □$$

Hence the root locus for a system with bounded realisation behaves in much the same way as the classical case if we consider the loci of generalised poles to correspond to the loci of ordinary poles in the latter case. Before giving some examples of this theory it is of interest, in view of corollary 7.3.6, to determine under which conditions a system has a simple spectral representation. It turns out that for a system with a bounded realisation which has algebraic singularities this is always possible. In order to prove this result we recall the theory of Rota (1960) which states that the left-shift operator on a suitable Hilbert space is a 'universal model' for any bounded operator whose spectrum is contained in the unit disc. In fact, if A is a bounded operator with $\sup |\sigma(A)| \leq 1$ defined on a Hilbert space H, then we define the new Hilbert space

$$H^\infty = \bigoplus_{k=1}^{\infty} H$$

which is the space of sequences $\mathbf{x} = (x_1,x_2,\ldots)$of elements x_i of H such that $\Sigma_{k=1}^{\infty}\|x_k\|^2 < \infty$ with inner product

$$\langle \boldsymbol{x},\boldsymbol{y}\rangle = \sum_{k=1}^{\infty} \langle x_k,y_k\rangle_H,\ x_k,y_k \in H$$

where

$$\mathbf{x} = (x_1,x_2,\ldots), \boldsymbol{y} = (y_1,y_2,\ldots)$$

Define the *left-shift* U_l on H^∞ by

$$U_l(x_1,x_2,x_3,\ldots) = (x_2,x_3,\ldots)$$

and the operator $I_A:H \rightarrow H$ by

$$I_A x = (x,Ax,A^2x,\ldots)$$

Then I_A is clearly invertible and

$$I_A^{-1} U_l I_A x = Ax$$

so that A is similar to U_l.

Consider again the general spectral decomposition of a system as in (7.15) and suppose that a particular projected subsystem defined by A_1,b_1,c_1 (say) has transfer function $G_1(s) = c_1R(s;A_1)b_1$. Assume that $\sigma(A_1)$ contains several generalised poles (neglecting 'finite-dimensional' singularities) as shown in Fig. 7.1. Suppose that the term $\Phi_1(s)$ in (7.20*a*) corresponds to the pole p_1. Then we can draw a Jordan curve γ in $\sigma(A_1)$ containing p_1 and excluding all other generalised poles. Since γ is a Jordan curve, the Riemann mapping theorem (Rudin, 1974) implies that there exists a conformal map f taking γ into the unit circle in the $f(s)$-plane. Now, f^{-1} is analytic so we can define $A_{1,1} \triangleq f^{-1}(U_l^{H_1})$ where $U_l^{H_1}$ is the left-shift operator on H_1. Then $A_{1,1}$ has spectrum consisting of γ and its interior (by the spectral mapping theorem). However, $A_{1,1}$ has the model U_l on H^∞; i.e.

$$A_{1,1} = I_{A_{1,1}}^{-1} U_l I_{A_{1,1}}$$

and we can realise $(\Phi_1(s))^{-1}$ on H^∞, using the results of Chapter 5, in terms of the triple (U_l,b,c), say. Hence, we obtain the system

$$\dot{\xi} = U_l\xi + bv$$

$$\dot{y} = c\xi$$

on H^∞. Changing variables to $\Xi = I_{A_{1,1}}^{-1}\xi$, we obtain the system

$$\dot{\Xi} = I_{A_{1,1}}^{-1} U_l I_{A_{1,1}} \Xi + I_{A_{1,1}}^{-1} bv$$

$$y = cI_{A_{1,1}}\Xi$$

or

$$\dot{\Xi} = A_{1,1}\Xi + b'v$$
$$y = c'\Xi$$

where

$$b' = I_{A_{1,1}}^{-1}b,\ c' = cI_{A_{1,1}}$$

The latter system is defined on H and has transfer function $(\Phi_1(s))^{-1}$ with spectrum $\gamma+$ interior of γ. This proves the above result concerning simple spectral representations.

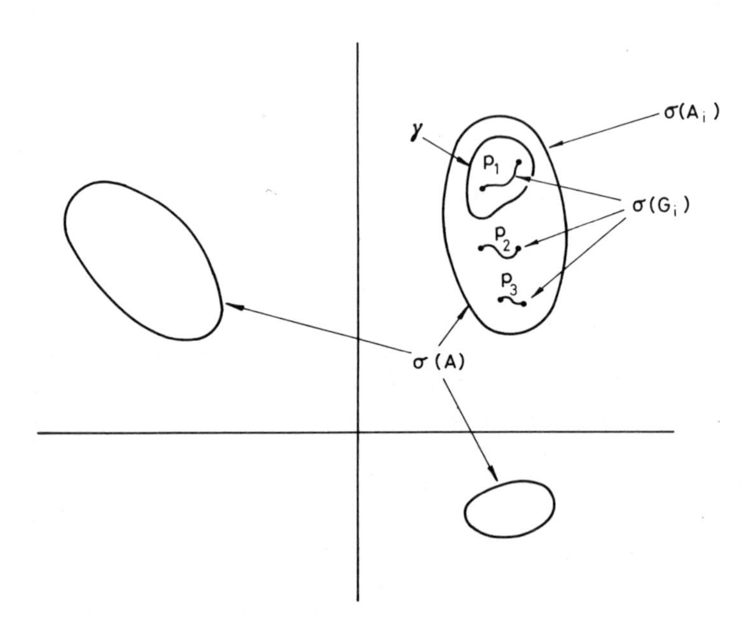

Fig. 7.1 *Typical spectral decomposition*

Returning to the general root locus theory for a system with algebraic singularities, we can show, by an elementary spectral continuity argument, that generalised poles remain generalised poles as k increases, although a generalised pole may become degenerate for some k and is, of course, then a classical pole. Combining this statement with corollary 7.3.8, we see that the generalised poles behave on the root locus in essentially the same way as classical poles for finite dimensional systems and explains why we have been concerned with connected components of $\sigma(G)$.

Before giving some examples we shall end this discussion of root locus theory with the important related pole assignment problem. For a system with bounded realisation we consider the *generalised pole assignment problem*; namely, can we place the generalised poles of such a system arbitrarily? The next result shows that this problem is approximately soluble.

Theorem 7.3.9
The generalised pole assignment problem is approximately soluble for the system (7.14) in the sense that there exists a sequence of finite dimensional systems $S_{0i} = (A_{0i}, b_{0i}, c_{0i})$ approximating $G_0(s)$, with transfer functions $G_{0i}(s)$, $1 \leqslant i < \infty$, such that, if $G_d(s)$ is the desired closed-loop transfer function then we can assign the poles of S_{0i} with state feedback so that

$$\overline{\bigcup_{i=1}^{\infty} \sigma(G_i)} = \sigma(G_d)$$

i.e., the union of the poles of all the systems S_i is dense in $\sigma(G_d)$. Here, S_i is the closed-loop system of S_{0i} and G_i is its transfer function.

Proof: This result follows easily from Runge's theorem. In fact, by the latter result, we can find a sequence of transfer functions $G_{0i}(s)$ of finite dimensional systems such that $G_{0i}(s) \rightarrow G_0(s)$ uniformly on compact sets of $\mathscr{C} \backslash \sigma(G_0)$. Similarly, we can find a sequence $G_i(s)$ such that $G_i(s) \rightarrow G_d(s)$ uniformly on compact subsets of $\mathscr{C} \backslash \sigma(G_d)$. Without loss of generality we assume that G_{0i} and G_i each have i poles (counted with multiplicity). By the classical result, we can assign the poles of S_{0i} arbitrarily; hence, we assign them to $\sigma(G_i)$ and a simple argument completes the proof. □

Example 7.3.10
In our first example, we shall consider the transfer function

$$G_0(s) = \frac{s+1}{\sqrt{(s^2+1)}}$$

with a classical zero at $s = -1$ and a generalised pole $(\pm i)$. Note that G_0 does not have a bounded (A, b, c) realisation but is the simplest nontrivial example which will bring out some important properties of irrational transfer functions.

The root locus is given by

$$1 + k \,.\, \frac{(s+1)}{\sqrt{(s^2+1)}} = 0$$

or

$$s^2 + 1 = k^2(s+1)^2$$

(Note that in squaring the above equation we have 'lost' the cut between $+i$ and $-i$, but we must bear in mind that it is present.) Hence,

$$s^2 - \frac{2k^2}{1-k^2}s + 1 = 0$$

and so

$$s = \frac{k^2 \pm \surd(2k^2 - 1)}{1 - k^2}$$

When $k = 0$ the root locus starts at $s = \pm i$ and has complex roots until $k = 1/\surd 2$ when the roots become real at the point $s = 1$. If $k \in [0, 1/\surd 2]$ we obtain the locus of a semicircle. If $k \in [1/\surd 2, 1)$ then we obtain two branches; one tends to $s = 0$ as $k \to 1$ and one tends to ∞ as $k \to 1$. However, as the theory predicts, the pole does not remain at ∞ but is attracted by the zero at $s = -1$. The locus is shown in Fig. 7.2.

Notice that for particular values of k, a generalised closed-loop pole is shown as (s_1,s_2) joined by a cut. When $k = 1/\surd 2$ the pair (s_1,s_2) coalesces at $s = 1$, and becomes a degenerate (or classical) pole, and when $k = 1$, s_2 is at ∞. The pole pair (s_1,s_2) together with the cuts can be shown to greater effect on the Riemann Sphere as in Fig. 7.3.

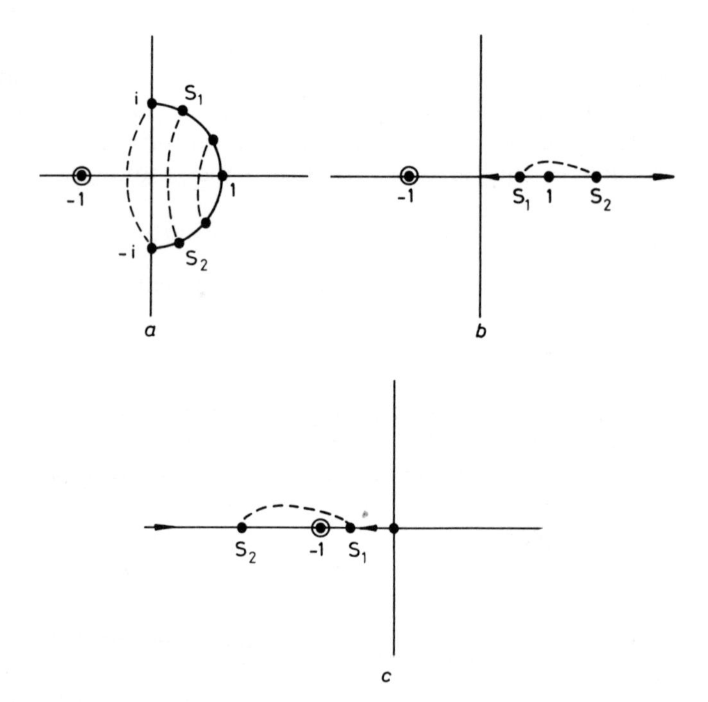

Fig. 7.2 *Root locus for example 7.3.10*
a $0 < k < 1/\surd 2$
b $1/\surd 2 \leqslant k < 1$
c $k > 1$

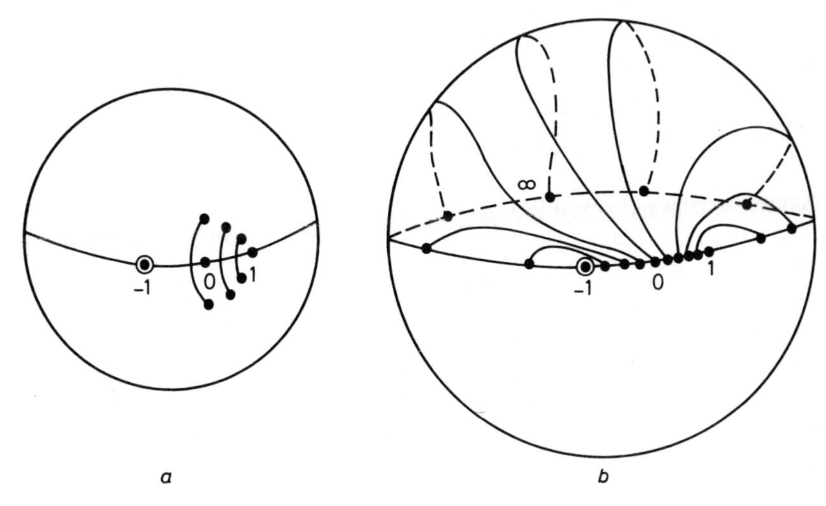

Fig. 7.3 *Root locus for example 7.3.10 displayed on the Riemann sphere*
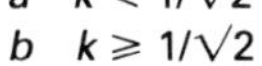
a $k < 1/\sqrt{2}$
b $k \geqslant 1/\sqrt{2}$

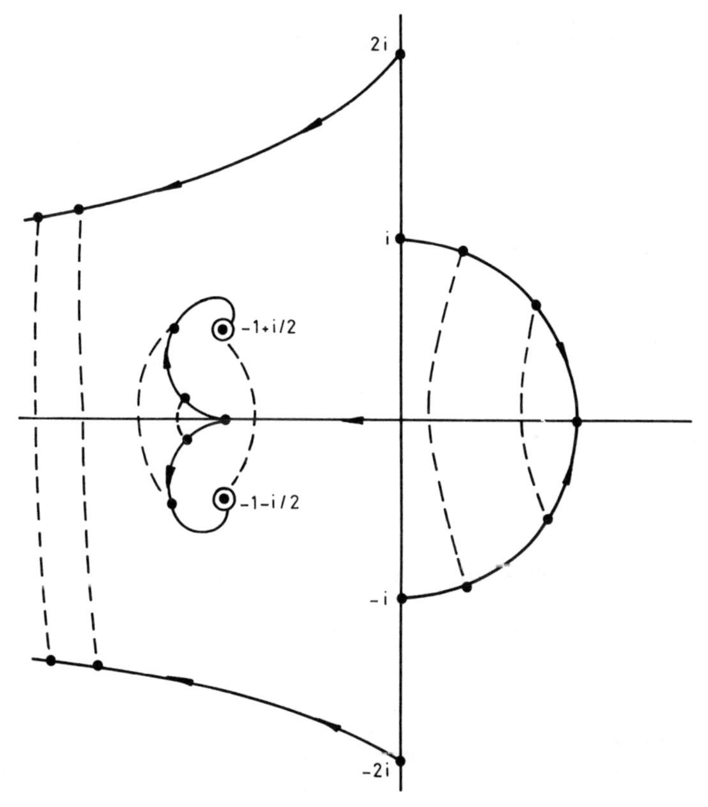

Fig. 7.4 *Root locus for example 7.3.11*

Example 7.3.11
Consider the transfer function

$$G_0(s) = \frac{\sqrt{((s+1)^2 - 1/4)}}{\sqrt{(s^2+4)}\sqrt{(s^2+1)}} \tag{7.22}$$

having two generalised poles at $(\pm i)$, $(\pm 2i)$ and a generalised zero at $(-1 \pm i/2)$. The root locus can be generated by computer and is shown in Fig. 7.4. As the theory shows, the generalised zero attracts one of the generalised poles and the other pole diverges to inifinity.

Before completing this section we shall consider briefly what may happen to the root locus of a system which is not single-input single-output. In fact we shall examine the approximation to the delay system in example 3.5.4. First, however, we must generalise the notion of transfer function and the root locus. Consider, then, the system

$$\begin{aligned} \dot{x} &= Ax + Bu \\ y &= Cx \end{aligned} \tag{7.23}$$

where, for simplicity, A,B,C will be assumed to be bounded operators on a Hilbert space H, and suppose that the control is a feedback of y. Then,

$$\dot{x} = Ax + B(v - ky)$$

$$y = Cx$$

By Laplace transformation,

$$sX(s) = AX(s) + B(V(s) - kY(s))$$

$$Y(s) = CX(s)$$

and so if $s \notin \sigma(A)$, we have

$$X(s) = R(A;s)B(V(s) - kY(s))$$

Hence,

$$\begin{aligned} Y(s) &= CR(A;s)B(V(s) - kY(s)) \\ &= G_0(s)(V(s) - kY(s)) \end{aligned}$$

where $G_0(s) \triangleq CR(A;s)B$ is the open-loop transfer function, and so

$$(I + kG_0(s))Y(s) = G_0(s)V(s)$$

If $0 \notin \sigma(G_0(s))$, then

$$G_0^{-1}(s)(I + kG_0(s))Y(s) = V(s) \tag{7.24}$$

We can solve (7.24) for Y in terms of V for any s for which

$$0 \notin \sigma(G_0^{-1}(s)(I + kG_0(s)))$$

Let $F(s) = G_0^{-1}(s)(I + kG_0(s))$, which exists as a bounded operator-valued analytic function of s for all s such that $0 \notin \sigma(G_0(s))$ and $s \notin \sigma(A)$. Denote the set of all such values s by Ω. Then, by the spectral mapping theorem,

$$\lambda \in \sigma(G_0(s)) \text{ iff } \frac{1}{\lambda}(1 + k\lambda) \in \sigma(F(s))$$

Let

$$\Gamma(k) = \{s \in \mathscr{C}: 0 \notin \sigma(F(s))\}$$

and let

$$\Gamma = \bigcup_{k\in[0,\infty)} \Gamma(k)$$

Then $\mathscr{C}\backslash\Gamma$ is the *root locus* (or spectral locus) of the system. Thus, if $s \in \Gamma(k)$, then

$$Y(s) = F^{-1}(s)V(s)$$

and $F^{-1}(s)$ is an analytic bounded operator-valued function on $\Gamma(k)$. Note that $F(s)$ was defined only for $s \notin \Omega$, although we may have $\Gamma(k)\backslash\Omega \neq \phi$; this means that $F^{-1}(s)$ is analytically continuable into $\Gamma(k)$.

Consider now the system

$$\begin{aligned} \dot{x} &= Ax + u \\ y &= x \end{aligned} \tag{7.25}$$

where A is the left-shift operator and we have taken $B = C = I$ for simplicity.

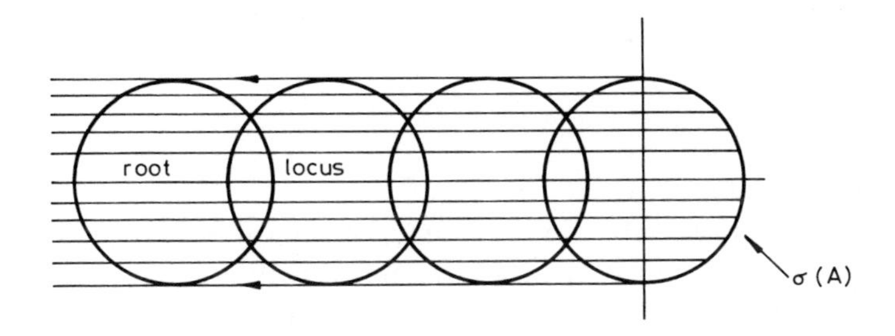

Fig. 7.5 *Root locus for the system (7.25)*

This is a simple model of the delay equation derived in example 3.5.4. Then,

$$G_0(s) = (sI - A)^{-1} = R(s;A),\ s \notin \sigma(A)$$

However,

$$\mu \in \sigma(A) \text{ iff } \frac{1}{s - \mu} \in \sigma(R(s;A))\ (s \neq \mu)$$

and so

$$\mu \in \sigma(A) \text{ iff } (s - \mu)\left(1 + \frac{k}{s - \mu}\right) \in \sigma(F(s)) \; (s \neq \mu)$$

(by the spectral mapping theorem). Hence, for each $k \geqslant 0$,

$$0 \in \sigma(F(s)) \text{ iff } s = \mu - k$$

for some $\mu \in \sigma(A)$. It then follows that the root locus of this system is as shown in Fig. 7.5.

To give a less trivial example, consider the system

$$\begin{pmatrix} \dot{x}_1 \\ \dot{x}_2 \end{pmatrix} = \begin{pmatrix} A & I \\ 0 & A - 3I \end{pmatrix} \begin{pmatrix} x_1 \\ x_2 \end{pmatrix} + \begin{pmatrix} 0 \\ I \end{pmatrix} u \tag{7.26}$$

where $x_1, x_2 \in l^2$ and A is again the left-shift operator. Then,

$$R\left\{s; \begin{pmatrix} A & I \\ 0 & A - 3I \end{pmatrix}\right\} = \begin{pmatrix} (sI - A)^{-1} & (sI - A)^{-1}(sI - A + 3I)^{-1} \\ 0 & (sI - A + 3I)^{-1} \end{pmatrix}$$

and so the transfer function of this system is

$$G_0(s) = (sI - A)^{-1}(sI - A + 3I)^{-1}$$

and by the spectral mapping theorem

$$\lambda \in \sigma(A) \text{ iff } \frac{1}{s - \lambda} \cdot \frac{1}{s - \lambda + 3} \in \sigma(G_0(s)), \; s \neq \lambda, \lambda - 3$$

Hence, the root locus is given by those values s which satisfy

$$1 + k \frac{1}{s - \lambda} \cdot \frac{1}{s - \lambda + 3} = 0 \text{ for } \lambda \in \sigma(A)$$

and is shown in Fig. 7.6.

The above theory can be used in a rather obvious way to generalise classical compensation theory to systems with bounded realisations. For systems with unbounded operators which have compact resolvent (in particular the heat equation), Pohjolainen (1982) has considered robust multivariable PID controllers. However, this work depends on the simplicity of the operators concerned and a more general theory for less well behaved operators is necessary.

7.4 Conclusion

In this book we have studied the state space and s-domain approaches to infinite dimensional systems theory. The linear state space theory may be said

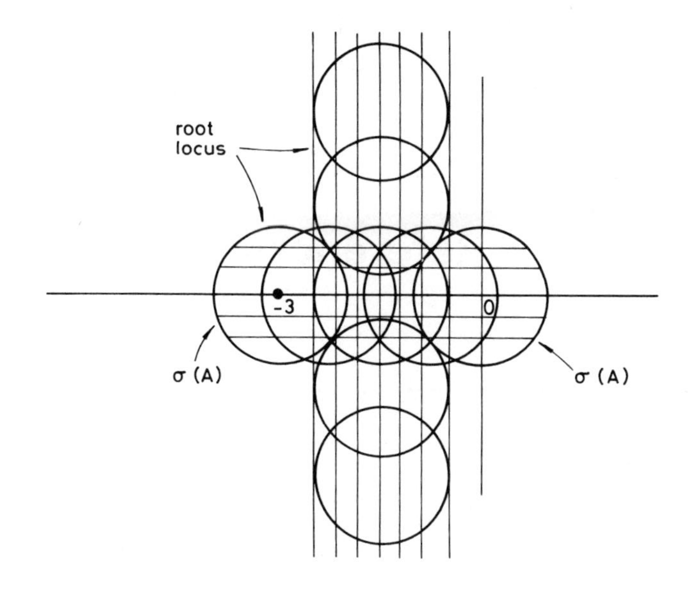

Fig. 7.6 *Root locus for the system (7.26)*

to be fairly well understood. However, the s-domain methods, particularly for general unbounded operators, remains an open problem. Moreover, the theory of control of nonlinear distributed systems has received little attention and will provide many interesting problems for the future, for example, in realisation theory and controllability questions.

References

ADAMS, R. A. (1975): 'Sobolev Spaces' (Academic Press)

AGMON, S. (1965): 'Lectures on elliptic boundary value problems' (Van Nostrand)

ARONSON, D. G. (1978): 'A comparison method for stability analysis of nonlinear parabolic problems', *SIAM Review*, **20**, pp. 245–264

BANKS, S. P. (1980): 'Guidance laws and receding horizon optimal control', Symposium on math. syst. theory, Warwick University

BANKS, S. P. (1981*a*): 'An abstract setting for nonlinear distributed multi-pass processes', *Int. J. Sys. Sci.*, **12**, pp. 1287–1301

BANKS, S. P. (1981*b*): 'The circle theorem for nonlinear parabolic systems', *Int. J. Control*, **34**, pp. 843–851

BANKS, S. P. (1981*c*): 'On the optimal control of residual stresses in high temperature materials', Research report No. 149, University of Sheffield

BANKS, S. P. (1982): 'Controllability, optimal control and receding horizon control of distributed multi-pass processes', *Int. J. Syst. Sci.*, **13**, pp. 1289–1311

BANKS, S. P. (1983): 'The receding horizon principle for distributed systems', *J. Franklin Inst.* **315**, pp. 435–451.

BANKS, S. P., and ABBASI-GHELMANSARAI (1983*a*): 'Delay equations, the left-shift operator and the infinite dimensional root locus', *Int. J. Control*, **37**, pp. 235–253

BANKS, S. P., and ABBASI-GHELMANSARAI, F. (1983*b*): 'Realisation theory and the infinite dimensional root locus', *Int. J. Control* (in press)

BANKS, S. P. and COLLINGWOOD, P. C. (1979): 'Stability of non-linearly interconnected systems and the small gain theorem', *Int. J. Control*, **30**, pp. 901–916

BANKS, S. P., and MOUSAVI-KHALKHALI, S. A. (1982): 'On optimal control of plasma confinement', *Int. J. Syst. Sci.*, **13**, pp. 757–771

BARAS, J. S. and BROCKETT, R. W. (1975): 'H^2-functions and infinite-dimensional realisation theory', *SIAM J. Control*, **13**, pp. 221–241

BARBU, V. (1976): 'Nonlinear semigroups and differential equations in Banach space' (Noordhoff)

BARBU, V. (1981): 'Necessary conditions for distributed control problems governed by parabolic variational inequalities', *SIAM. J. Control*, **19**, pp. 64–86

BODE, H. W. (1954): 'Network analysis and feedback amplifier design' (Van Nostrand)

BOYD, T. J. M., and SANDERSON, J. J. (1969): 'Plasma dynamics' (Nelson)

BROCKETT, R. W., and FUHRMANN, P. A. (1976): 'Normal symmetric dynamical systems', *SIAM J. Control*, **14**, pp. 107–119

BUIS, G. R. (1968): 'Lyapunov stability for partial differential equations', Part 1, NASA CR-1100

CIRINA, M. (1969): 'Boundary controllability of nonlinear hyperbolic systems', *SIAM J. Control*, **7**, pp. 198–212

CRANDALL, M. G., and RABINOWITZ, P. H. (1977): 'The Hopf bifurcation theorem in infinite dimensions', *Arch. fur Rat. Mech. Anal.*, **67**, pp. 53–72.

CURTAIN, R. F., and PRITCHARD, A. J. (1978): 'Infinite dimensional linear systems theory' (Springer-Verlag)

DATKO, R. (1970): 'Extending a theorem of A. M. Lyapunov to Hilbert space', *J. Math. Anal. Appl.*, **32**, pp. 610–616

DELFOUR, M. C., and MITTER, S. K. (1972): 'Controllability, observability and optimal feedback control of affine hereditary differential systems', *SIAM J. Control*, **10**, pp. 298–328

DESOER, C. A., and VIDYASAGAR, M. (1975): 'Feedback systems: input-output properties' (Academic Press)

DESOER, C. A., and WANG, Y-T. (1980): 'On the generalised Nyquist stability criterion', *Trans. IEEE*, **AC-25**, pp. 187–196

DOLECKI, S., and RUSSELL, D. L. (1977): 'A general theory of observation and control', *SIAM J. Control*, **15**, pp. 185–227

DUNFORD, N., and SCHWARTZ, J. T. (1959): 'Linear operators', Vol. I, (1963) Vol. II, (Interscience)

EDWARDS, J. B., and OWENS, D. H. (1981): 'Analysis and control of multipass processes' (Research Studies Press, Wiley)

FATTORINI, H. O. (1964): 'Time-optimal control of solutions of operational differential equations', *SIAM J. Control*, **2**, pp. 54–59

FATTORINI, H. O. (1975*a*): 'Boundary control of temperature distributions in a parallelipipedon', *SIAM J. Control*, **12**, pp. 1–13

FATTORINI, H. O. (1975*b*): 'Local controllability of a nonlinear wave equation', *Math. Systems Theory*, **9**, pp. 30–45

FATTORINI, H. O. (1977): 'The time optimal problem for distributed control of systems described by the wave equation', *in* AZIZ, A. K., WINGATE, J. W., and BALAS, M. J. (eds.) (Academic Press)

FATTORINI, H. O., and RUSSELL, D. L. (1971): 'Exact controllability theorems for linear parabolic equations in one space dimension', *Arch. fur Rat. Mech. Anal.*, **43**, pp. 272–292

FEINTUCH, A., and ROSENFELD, M. (1978): 'On pole assignment for a class of infinite dimensional linear systems', *SIAM J. Control*, **16**, pp. 270–276

FREEDMAN, M. I., FALB, P. L., and ZAMES, G. (1969): 'A Hilbert space stability theory over locally compact Abelian groups', *SIAM J. Control*, **7**, pp. 479–495

FRIEDMAN, A. (1964): 'Partial differential equations of parabolic type' (Prentice-Hall)

FRIEDMAN, A. (1969): 'Partial differential equations' (Holt, Rinehart, Winston)

GIBSON, J. S. (1979): 'The Riccati integral equations for optimal control problems on Hilbert spaces', *SIAM. J. Control*, **17**, pp. 537–565

GIRSANOV, I. V. (1972): 'Lectures on the mathematical theory of extremum problems' (Springer-Verlag)

GUPTA, S. C., and HASDORFF, L. (1970): 'Fundamentals of automatic control' (Wiley)

HAHN, W. (1956): 'Theory and applications of Lyapunov's direct method' (Prentice-Hall)

HALANAY, A. (1966): 'Differential equations; stability, oscillation, time lags' (Academic Press)

HALE, J. K. (1969): 'Dynamical systems and stability', *J. Math. Anal. Appl.*, **26**, pp. 39–59

HAMATSUKA, T., MO'OMEN, A-A, and AKASHI, H. (1981): 'On pole assignment and stabilisation for the heat equation', *Mem. Fac. Eng. Kyoto Univ. (Japan)*, **43**, pp. 319–327

HELTON, J. W. (1976): 'Systems with infinite dimensional state space: The Hilbert space approach', *Proc. IEEE*, **64**, pp. 145–160

HENRY, D. (1981): 'Geometric theory of semilinear parabolic equations' (Springer-Verlag)

HILLE, E., and PHILLIPS, R. S. (1957): 'Functional analysis and semi-groups', *Amer. Math. Soc. Coll. Publ.*, **31**

HOLMES, P., and MARSDEN, J. (1981): 'A partial differential equation with infinitely many periodic orbits: chaotic oscillations of a forced beam', *Arch. fur Rat. Mech. Anal.*, **76**, pp. 135–165

HORVATH, J. (1966): 'Topological vector spaces and distributions' (Addison-Wesley)
KASTENBERG, W. E. (1969): 'Stability analysis of nonlinear space dependent reactor kinetics', *Adv. in Nucl. Sci. and Tech.*, **5**
KATO, T. (1976): 'Perturbation theory for linear operators' (Springer-Verlag)
KIRCHGASSNER, K., and KIELHOFER, H. (1973): 'Stability and bifurcation in fluid mechanics', *Rocky Mtn. Math. J.*, **3**, pp. 275–318
KUSAKABE, T., and MIHARA, Y. (1980): 'Control of residual stress in hot-rolled H-shapes', *Trans. ISIJ*, **20**, pp. 518–525
LASALLE, J. P., and LEFSCHETZ, S. (1961): 'Stability by Lyapunov's direct method with applications' (Academic Press)
LEE, E. B., and MARKUS, L. (1967): 'Foundations of optimal control theory' (Wiley)
LIONS, J. L. (1971): 'Optimal control of systems governed by partial differential equations' (Springer-Verlag)
LIONS, J. L., and MAGENES, E. (1972): 'Nonhomogeneous boundary value problems', Vol. I (Springer-Verlag)
LYAPUNOV, A. A. (1949): 'Problème général de la stabilité du mouvement' (Princeton University Press)
NYQUIST, H. (1932): 'Regeneration theory', *Bell. Sys. Tech. J.*, **11**, pp. 126–147
OWENS, D. H. (1978): 'Feedback and multivariable systems' (Peter Peregrinus)
PAO, C-V. (1968): 'Stability of nonlinear operational differential equations in Hilbert space', Ph.D. Thesis, University of Pittsburgh
POHJOLAINEN, S. (1981): 'Computation of transmission zeros for distributed parameter systems', *Int. J. Control*, **33**, pp. 199–212
POHJOLAINEN, S. (1982): 'Robust multivariable PI-controller for infinite dimensional systems', *IEEE Trans. Aut. Contr.*, **AC-27**, pp. 17–30
PONTRYAGIN, L. S., BOLTYANSKII, V. G., GAMKRELIDZE, R. V., and MISHCHENKO, E. F. (1962): 'The mathematical theory of optimal processes' (Interscience, Wiley)
POSTLETHWAITE, I., and MACFARLANE, A. G. J. (1979): 'A complex variable approach to the analysis of linear multivariable systems' (Springer-Verlag)
PRITCHARD, A. J., and ZABCZYK, J. (1977): 'Stability and stabilisability of infinite dimensional systems', CTC Report No. 70, Warwick University
RIESZ, F., and SZ-NAGY, B. (1955): 'Functional analysis' (Academic Press)
ROSENBROCK, H. H. (1974): 'Computer aided control system design' (Academic Press)
ROTA, G-C. (1960): 'On models for linear operators', *Comm. on Pure and Appl. Maths.*, **13**, pp. 469–472
RUDIN, W. (1974): 'Real and complex analysis' (McGraw-Hill)
RUSSELL, D. L. (1974): 'Exact boundary value controllability theory for wave and heat equations in star-complemented regions', *in* Roxin, Liu and Sternberg (eds.) 'Diff. games and control theory' (Dekker)
RUSSELL, D. L. (1978*a*): 'Controllability and stabilisability theory for linear partial differential equations: recent progress and open questions', *SIAM Review*, **20**, pp. 639–739
RUSSELL, D. L. (1978*b*): 'Canonical forms and spectral determination for a class of hyperbolic distributed parameter control systems', *J. Math. Anal. Appl.*, **62**, pp. 186–225
SANDBERG, I. W. (1964): 'On the L_2-boundedness of solutions of non-linear functional equations', *Bell Syst. Tech. J.*, **43**, pp. 1581–1599
SHAW, L. (1979): 'Nonlinear control of linear multivariable systems via state-dependent feedback gains', *Trans. IEEE*, **AC-24**, pp. 108–112
SLEMROD, M. (1974): 'A note on complete controllability and stabilisability for linear control systems in Hilbert space', *SIAM J. Control*, **12**, pp. 500–508
SLEMROD, M. (1976): 'Asymptotic behaviour of C_0 semigroups as determined by the spectrum of the operator', *Indiana Math. J.*, **25**, pp. 783–792
TAYLOR, A. E. (1958): 'Functional analysis' (Wiley)

TREVES, F. (1967): 'Topological vector spaces, distributions and kernels' (Academic Press)
TRIGGIANI, R. (1975): 'On the stabilisability problem in Banach space', *J. Math. Anal. Appl.*, **52**, pp. 383–403
WALKER, J. A. (1976): 'On the application of Lyapunov's direct method to linear dynamical systems', *J. Math. Anal. Appl.*, **53**, pp. 187–220
WASHBURN, D. (1979): 'A bound on the boundary input map for parabolic equations with applications to time optimal control', *SIAM J. Control*, **17**, pp. 652–671
WHITTAKER, E. T., and WATSON, G. N. (1965): 'A course of modern analysis' (Cambridge University Press)
WOLOVICH, W. A. (1974): 'Linear multivariable systems' (Springer-Verlag)
WONHAM, W. M. (1979): 'Linear multivariable control: a geometric approach' (Springer-Verlag)
YOSHIZAWA, T. (1966): 'Stability theory by Lyapunov's second method' (Math. Soc. of Japan)
YOSIDA, K. (1974): 'Functional analysis' (Springer-Verlag)
ZABCZYK, J. (1976): 'Remarks on the algebraic Riccati equation in Hilbert space', *Appl. Math. and Opt.*, **3**, pp. 383–403
ZAMES, G. (1966): 'On the input-output stability of time-varying nonlinear feedback systems', *Trans. IEEE*, **AC-11**, Part I, pp. 228–238; Part II, pp. 465–476

Index